NUMERICAL PETROLOGY

Statistical Interpretation of Geochemical Data

Series

Developments in Petrology

Developments in Petrology 8

NUMERICAL PETROLOGY

Statistical Interpretation of Geochemical Data

R.W. LE MAITRE

Department of Geology, University of Melbourne, Melbourne, Parkville, Vic. 3052 Australia

ELSEVIER SCIENTIFIC PUBLISHING COMPANY
AMSTERDAM — OXFORD — NEW YORK   1982

ELSEVIER SCIENTIFIC PUBLISHING COMPANY
Molenwerf 1
P.O. Box 211, 1000 AE Amsterdam, The Netherlands

Distributors for the United States and Canada:

ELSEVIER SCIENCE PUBLISHING COMPANY INC.
52, Vanderbilt Avenue
New York, N.Y. 10017

ISBN 0-444-42098-3 (Vol. 8)
ISBN 0-444-41562-9 (Series)

Printed in The Netherlands

PREFACE

Most scientific data are of a multivariate nature, and petrology is no exception. The more we delve into the details of petrological processes the more it becomes obvious that there are many factors involved which have not been taken into account. This is particularly true of petrology and geochemistry as more and more elements are commonly being analysed on a routine basis - aided and abetted by sophisticated analytical equipment which makes the task that much easier. As a result there has been an enormous explosion of petrological data over the last two decades, in both the number of analyses and in the number of elements determined.

In order to process this data efficiently, petrologists will have to become more numerate. Just as they had to master the complicated electronic equipment with the advent of, for example, the electron microprobe in the early 60's, so they will have to learn statistical and computing techniques in the next decade, if they wish to keep in the forefront of research.

The main objective of this book, therefore, is to try to present under one cover some of the techniques that are likely to be useful to the average petrologist, together with some of the thoughts that have occurred during many years of research and teaching in the field of what I have loosely called "numerical petrology". Although most of the contents of the book can be found in a variety of statistical publications, some is new material. In particular the sections on generalized petrological mixing models and their errors in Chapter 6; on dealing with the problems of closed data and singular matrices in multiple discriminant analysis and classification procedure in Chapters 9 and 10, respectively; and on the propagation of errors in mineral and normative recalculations in Chapter 11. Several of the Tables in the Appendix are also new and have been compiled with the numerate petrologist in mind. In particular the theoretical chemical compositions of most of the common rock-forming minerals given in Table 13 which, among other things, can be useful for mixing model calculations, and Table 14 which gives the common silicates in sorted order of their chemical composition.

As I have assumed that the reader has a background in petrology rather than statistics and mathematics, I have included many diagrams and a large number of worked numerical examples all from the field of petrology, as I believe that this is what petrologists need and require to become familiar with some of the techniques that they will have to learn. For similar reasons, most of the complicated mathematics, that is not absolutely essential to the understanding of the problem in hand, has been omitted and references given to where it can be obtained, if required. This includes the theory of matrices which, of course, is given in innumerable texts.

Throughout the book much emphasis has been placed upon the

interpretation of the results, as most petrologists do not write their own computer programs to process their data, but rely upon other peoples. It is, however, essential that they have a clear understanding of how their results were obtained, so that they can interpret them correctly. While this does not mean that all petrologists must be fully conversant with computer programming, it is highly recommended that they have a working knowledge of it so that, at least, they can make a few minor modifications to existing programs. It is then but a short step to writing a program of their own!

The reader may also be interested in the production of this book, as without computers parts of it would never have been produced. The book was typed directly into a word processor, using the WORD 11 system on the University of Melbourne PDP 11/70. All the data for the numerical examples and Tables were prepared in the required format on the University of Melbourne CDC Cyber 73, transferred to the PDP 11/70 and put directly into WORD 11, so that nearly all of the Tables are "untouched by human hand", so to speak. This greatly minimized the problem of typographical errors in the Tables, which are always difficult to detect. However, as a result of the calculations being performed with far more significant figures than are printed, some of the tabulated data will not reproduce *exactly* the results shown, due to rounding-off errors. The photo-ready text was then printed on a Diablo 630 printer, using a program that enables up to four different print-wheels to be used for each page of text. However, due to the lack of print-wheels with suitable typefaces, the Greek letters, bold-faced characters and summation signs had to be inserted by hand. This is also why the super- and sub-scripted numerals are the same size as the normal numerals.

Finally, I would like to extend my sincere thanks to all those who have assisted and put up with me during the production of this book. In particular to my wife for constant encouragement; to Carol Sugden for secretarial assistance and advice; to Don Campbell and Mary Hutchison for drafting the diagrams; to Margaret Mitchell and Peter Hylands for invaluable advice on editorial and artistic matters; to the staff of the Computer Centre, especially Geoffrey Hudson and Jefferey McDonell, for much help in pushing WORD 11 to its absolute limits - and at times beyond; to F.H.C. Marriott for kind permission to use one of his diagrams, which is presented in a slightly modified form as Fig. 10.2; and to Felix Chayes, Aldo Cundari, Max Hey, Geoffrey Hill, Mike Horder, Bob Howie, Al Miesch, Felix Mutschler and Evan Williams for critically reviewing various parts of the book - their suggestions were most valuable. However, I must apologize to one or two of the reviewers for not incorporating *all* of their suggestions, as there are a few points on which we must agree to differ.

Melbourne, Australia.

April, 1982

R.W. LE MAITRE

TABLE OF CONTENTS

CHAPTER 4.   LINEAR RELATIONSHIPS BETWEEN TWO VARIABLES

CHAPTER 5.   THE GEOMETRY OF PETROLOGICAL DATA SPACE

CHAPTER 6.   MULTIPLE LINEAR REGRESSION AND PETROLOGICAL
MIXING MODELS

CHAPTER 7.   PRINCIPAL COMPONENTS ANALYSIS

CHAPTER 8.   FACTOR ANALYSIS

CHAPTER 9.   MULTIPLE DISCRIMINANT ANALYSIS

CHAPTER 10.   CLUSTER ANALYSIS AND CLASSIFICATION

CHAPTER 11. PROPAGATION OF ERRORS

CHAPTER 12. PETROLOGICAL DATA MANAGEMENT

CHAPTER 1

THE NATURE OF PETROLOGICAL DATA

1.1 INTRODUCTION

The petrologist of today not only has to be a master of the petrological microscope, but also has to be numerate to interpret the increasing amounts of data produced by modern analytical equipment. A clear understanding of the nature of this type of petrological data is, therefore, essential. Petrological data that can be quantified include both *discrete* and *continuous variables*. Discrete variables are those that can only have certain values e.g. the number of volcanoes per 100 square kilometers. The majority of petrological data are, however, continuous and can have any values, although many are also constrained to lie within certain limits, such as wt% (weight%) which must lie between 0 and 100. Some of these data also have the appearance of discrete variables as they are rounded-off to a specific number of decimal places, such as ppm (parts per million) and wt%, which are normally rounded to 0 and 2 decimal places, respectively. Many of the commonly used variables are also dimensionless, being expressed in terms of weight per unit weight or volume per unit volume, e.g. wt%, proportions, volume%, ppm and ppb (parts per billion).

Petrological variables can be further divided into primary and secondary variables. *Primary variables* are those that are measured by chemical analysis, observation or experiment e.g. wt%, ppm, density, etc., and should be the prime content of any data file. *Secondary variables* are those that are calculated from primary or other secondary variables e.g. CIPW normative values, iron-magnesium ratios etc. If execution time is cheap compared to mass storage, secondary variables should not be stored on a computer based data file, as they can always be calculated when required.

1.2 SOURCES OF ERROR IN CHEMICAL ANALYSES

When a petrologist analyses specimens of a granite and draws conclusions about the composition of the granite batholith from which they came, he has become involved with *statistical inference*. This is the process whereby quantitative statements are made about properties of a *population* (all possible specimens that could be collected from the granite batholith) from knowledge of data obtained from a *sample* (all the specimens that were analysed). Note that statisticians use the word "sample" in a slightly different sense to geologists. A "sample" to a statistician is a number of objects taken from a population, whereas a "sample" to a geologist is usually a single object or specimen. In the rest of this book the term "sample" will refer to the statistician's sample and the term "specimen" will be used for the geologist's "sample". A population can be of any size, from finite to infinite, and may be thought of as the entire body of objects of current interest. Generally speaking, however, it is not possible or practical to obtain all the objects, so that a sub-set called a sample is taken. As Snedecor & Cochran (1967, p.4) state, "It is the sample that we observe, but it is the population which we seek to know".

Once measurements have been made upon the sample, statements can then be made about the population. However, if the sample has been collected in a biased manner, e.g. by only collecting weathered specimens, then the statements will be applicable to a new *sub-population* of weathered specimens, from which the sample could have been randomly selected.

1.2.1 *Random sampling procedures*

To avoid bias the sample should be chosen at random, so that every object in the population has an equal chance of being collected, but as Koch & Link (1970, p.271) state, "The human tendency to collect oddities is revealed by the usual rock collection, which contains more strange than common rocks". This is true of many collections of volcanic rocks where dykes are often over represented.

Unfortunately, the chance of introducing bias into petrological sampling seems to be greater than in many other types of sampling as up to 5 distinct stages may be involved in obtaining the data:-
1. The analysed powder is a sample from a larger amount of crushed rock.
2. The crushed rock is a sample from a hand specimen.
3. The hand specimen chosen for analysis is a sample from all the specimens collected.
4. Each specimen collected is a sample from an outcrop.
5. Each outcrop selected for collection is a sample of the rock body under investigation.

Stages 3, 4 and 5, which are applicable to all geological collecting, are particularly prone to the inclusion of oddities and if the petrologist is seriously interested in the rock mass as a whole (the population), bias in these steps should be avoided wherever possible.

This can be done by using some kind of *random sampling* procedure, involving the use of random numbers. Such methods can be used for stages 2 to 5, or any situation where specimens or outcrops can be individually numbered. The selection of objects to be included in the sample is then done with reference to random numbers such as given in Table A1 in the Appendix. This contains 5,000 random numbers such that every integer from 0 to 9 has an equal chance of occurring in every location of the table. The numbers are arranged in blocks of 5x5 for ease of reading and each block and line is also numbered. To select numbers from a table of random numbers, choose a starting point at random (a stab of a pencil with the eyes shut will do) and then proceed to extract the numbers in either direction, horizontally or vertically.

To illustrate the procedure imagine we have 10 specimens of a basalt lava flow and we can only afford to analyse 3 of them. Which 3 should we choose to get an unbiased sample? The initial pencil jab may have landed on line 28 and block 3, which gives the string of numbers 92743. Reading from left to right we would then select specimens 9, 2 and 7 for analysis. Had the same number occurred twice we would ignore it and proceed to the next. This example is a simple one as there are 10 specimens and 10 digits in the table.

However, the number of specimens will generally not be 10, so how then do we proceed? If the number of specimens is less than 10, we simply ignore any digits that are too large, as all digits have an equal chance of occurring in the table. Therefore, if we only had 8 specimens to start with, specimens 2, 7 and 4 would have been chosen for analysis.

We can adopt a similar procedure if the number of specimens is greater than 10, by selecting the digits from Table A1 in pairs, if the number of specimens is less than 100, in triplets if the number of specimens is less than 1000, etc.. For example, if we have 17 specimens of the basalt flow and still wish to select 3, we would select *pairs* of numbers from Table A1 which, starting at line 28 and block 3 as before, would give the following sequence of numbers:-

92, 74, 36, 77, 39, 53, 56, 73, _02_, 39, 54, _12_, 98, 64, 96, 36, _12_, 71, 36, 35, _13_, 36, 17, 06, 27, ...

from which we would select 2, 12 and 13, as they are the first 3 different numbers less than 18. Note, however, that this is wasteful of numbers as we have had to extract 21 numbers before we found 3 less than 18.

An alternative strategy, which uses every number, is to divide each of these numbers by 17 (the number of specimens) and use the remainder, which will lie between 0 and 16, as the random number by which to select the specimens. This is often called the *modulo* function and is found on all computers and some electronic calculators. Using this method the 3 specimens chosen would have been 7 (17 divided into 92 goes 5 with 7 left over), 6 and 2.

Such a randomizing procedure should also be adopted for the order in which analyses are performed to minimize any bias that may be present in the analytical procedure.

A commonly used scheme for selecting outcrops from which to collect specimens, is to erect a grid over the collecting area and to then take specimens, either at all the grid points or at randomly chosen grid points.

This, however, is often impractical in areas of poor outcrop, unless one is prepared or able to drill for specimens. Alternatively, all the known outcrops can be numbered and then randomly selected for the collection of specimens.

Further information and other applications in a geological situation, can be found in Gy (1979), Koch & Link (1970), Miesch (1976a) and Till (1974). A detailed discussion of the theory of errors involved in geochemical sampling is given by Miesch (1967a).

1.2.2 Homogeneity of rock powders

Even though bias may be minimized by random sampling procedures in stages 3, 4 and 5, problems can still occur in the crushing and sampling involved in stages 1 and 2. For example, Kleeman (1967), using the binomial distribution, has suggested that if the number of grains in the rock powder being analysed is less than 10^6, the errors in the analysis may become significant. He, therefore, recommends that powders for analysis should be crushed to less than 120 mesh, or finer if the sample weight is less than 0.5 gms. For reference standards he advocates crushing to pass 230 mesh, but this conflicts with later work of Ridley et al. (1976), who show that the sampling errors do not decrease regularly with grain size, but that −140 to +200 mesh is the best grain size range for minimizing analytical errors. Their work was purely empirical and based on a crushed sample of 5030 gms. of a medium to coarse-grained granodiorite, from which they conclude that most of their sources of analytical error were attributable to heterogeneity in the powdered sample. They advocate the fusion of the powder to glass as the best homogenizing procedure.

Ingamells et al. (1972) and Ingamells (1974) use the binomial and normal distributions to determine the theoretical sampling errors in bimineralic mixtures, and give curves relating the relative errors to grain size and the proportion of the two minerals in the mixture. Although this may not seem particularly pertinent to actual rocks, it is applicable to many trace elements which are often concentrated in one particular mineral phase. Clanton & Fletcher (1976) also suggest that much of the variability in the analyses of lunar rocks may be due to the specimens being too small. Using a theoretical model based on the shape of the mineral grains in the rock, they use Monte Carlo methods to simulate the sampling errors and produce diagrams relating the rock texture to the sample size that must be taken to obtain a specified error in the oxides. Another theoretical approach is given by Moore (1979) who uses the Poisson distribution to derive expressions for the coefficient of variation (equation 2.5) of trace elements in granular materials and shows how they are related to particle size and density.

Much has also been written on this subject with particular reference to the USGS geochemical standards W-1 and G-1 (e.g. Chayes, 1970a; Vistelius, 1970; Flanagan, 1976c), but there appears to be no clear consensus as to the cause of the variations in the analyses of these standards. Results from a later set of standards designed to investigate this problem do, however, indicate that with proper procedures the analytical differences between powdered samples can, in most cases, be reduced to an insignificant level (Flanagan, 1976a). Engels & Ingamells (1977) also discuss the procedures for the preparation of homogeneous material for use as geochemical standards and also methods of reporting the analytical results. Although petrologists may not think it necessary to go to these lengths, they should be aware of the problems.

The problem, of course, will always be greater with trace elements, especially those which have a very high concentration in one minor mineral phase e.g. Zr in zircon, Au in native gold. Some procedures for dealing with these problems are discussed by Flanagan (1976a) and Koch & Link (1970).

1.2.3 Precision versus accuracy

All methods of chemical analysis are subject to *analytical errors* which can be thought of as consisting of two main components. One is the *statistical error* or *precision* which arises from purely random fluctuations in the analytical procedure, such as those encountered with all x-ray counting techniques. The other is the *systematic error* or *accuracy* whereby a certain bias is present in the results, as could happen if a badly determined calibration curve is used. Precision is, therefore, a measure of the reproducibility of a set of results and is not the same as accuracy, which is a measure of how close the results are to the true value. To give an analogy, if we have a ruler with an incorrectly engraved scale, we might be able to measure with precision, but the results would be inaccurate.

In practice it is difficult to separate these two types of error, but with certain comparative analytical techniques the precision can be partly controlled by the analyst. For example, both x-ray fluorescence and electron microprobe analysis are based on an equation of the type:-

$$\text{Amount in unknown} = \frac{\text{Counts in unknown}}{\text{Counts in standard}} \times \text{Amount in standard} \qquad (1.1)$$

Now although the two counts may be determined with great precision by counting for a long time, the relative error of the amount in the unknown can never be reduced below the relative error of the amount in the standard, due to the propagation of errors (see section 11.2.4). It is, therefore, important to know the analytical errors associated with the standard. But how can these be assessed?

With some standards the analytical errors can be regarded as negligible, e.g. with "pure" phases and compounds such as quartz or metallic elements. However, when this type of standard is used, large theoretical or empirical corrections often have to be applied, which introduces another source of error that is extremely difficult to quantify. Even if the standard has been prepared to a particular specification, uncertainties in its final composition can arise due to high temperature treatment, inhomogeneities or the development of reaction products (Shaw, 1969). It is for this reason that geologists often prefer to use natural standards, but the only true way to assess their analytical errors is to have them analysed many times by an absolute method such as wet chemical analysis.

To give some idea of the order of magnitude of these analytical errors, some data on the USGS standards G-1 and W-1 from Stevens & Niles (1960) are given in Table 1.1. The standard deviation of the means can be taken as the analytical errors of the oxides. The use that can be made of this information will be explained in section 1.3.1 and Chapter 2. It should be obvious, however, that the second decimal place for most major oxides has little meaning. Similar data can be found in Russell et al. (1972) and Steele & Hansen (1979) for major elements determined on the NIMROC reference samples, although the results are quoted as coefficients of variation (equation 2.5). Data such as these and those given in Table 1.1 should only be used as a guide and should not

TABLE 1.1

Data on the USGS standards G-1 and W-1 taken from Stevens & Niles (1960). The standard deviation of the mean was not given and has been calculated from their data.

Oxide	USGS standard	Number of analyses n	Mean $\bar{x}$	Standard deviation s	Standard deviation of mean $s/\sqrt{n}$
SiO_2	G-1	60	72.35	0.48	0.062
	W-1	60	52.40	0.33	0.043
TiO_2	G-1	60	0.26	0.04	0.005
	W-1	60	1.07	0.20	0.026
Al_2O_3	G-1	60	14.32	0.37	0.048
	W-1	60	15.11	0.63	0.081
Fe_2O_3	G-1	57	0.95	0.30	0.040
	W-1	58	1.62	0.71	0.093
FeO	G-1	57	0.99	0.11	0.015
	W-1	58	8.63	0.41	0.054
MgO	G-1	59	0.40	0.13	0.017
	W-1	59	6.58	0.35	0.046
CaO	G-1	59	1.40	0.12	0.016
	W-1	59	10.97	0.16	0.021
Na_2O	G-1	59	3.31	0.23	0.030
	W-1	58	2.07	0.20	0.026
K_2O	G-1	59	5.42	0.39	0.051
	W-1	58	0.67	0.13	0.017

be extrapolated to other rocks, as it is clear that the coefficient of variation increases markedly as the concentration of certain elements decreases (Rubeska,1977). Abbey (1977) and Ahrens (1977) also make some enlightened comments on the philosophy of geochemical "standards". Further information on these topics can be found in Shaw (1969).

1.3 FREQUENCY DISTRIBUTIONS AND PROBABILITY

If a population was completely uniform in its characteristics, only one object would need to be sampled to provide all the information about the population. However, nature does not work in this way. From experience we know that if we repeatedly select objects from a natural population and make measurements upon them, each will be slightly different. Thus every specimen of granite from a batholith will have a slightly different composition, even if there is absolutely no error involved in the analytical procedure.

In general the values will tend to cluster about a central value and their frequency of occurrence will decline away from this value. If the values are grouped together in classes, such that each class represents a particular range of values, a *histogram* can be produced showing the frequency of occurrence of each class. This is shown in Fig. 1.1 for the values of SiO_2 in 2578 analyses of andesites from the CLAIR data file as used by Le Maitre (1976a). It can be seen that the class containing most analyses has SiO_2 values ranging from 58 to 60, while analyses in other classes occur with decreasing frequency away from the centre. The frequency may be expressed either as the actual number of observations in the class, or as the relative frequency, which is the actual number in the class divided by the total number, or as a percentage frequency.

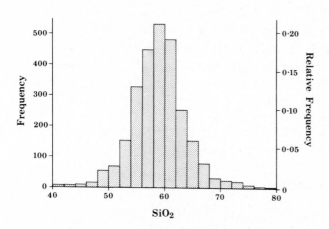

Fig. 1.1. Histogram of SiO_2 values of 2578 andesite analyses taken from the CLAIR data file. The class width is 2%. The vertical axis is marked in both absolute frequency and relative frequency.

We can then state that, if we choose an analysis at random from the population of 2578 andesites, there are 521 chances in 2578, or a probability of 0.202, of SiO_2 lying between 58 and 60 i.e. the relative frequencies are probabilities.

A histogram, such as Fig 1.1, is an *empirical frequency distribution* or, if scaled with relative frequencies so that the total area of all the rectangles is 1, is an *empirical probability distribution*. However, histograms suffer from the disadvantage that they are *discontinuous distributions* i.e. frequencies are known only for discrete ranges of values. In many instances, a *continuous distribution* in the form of a curve, which can be represented by a mathematical formula, is more convenient to use. Although a smooth curve could be drawn through the midpoints of the top of each rectangle in Fig. 1.1, its mathematical expression would still be in doubt. This is one reason why assumptions are often made about the nature of the population probability distribution. For example, it might be assumed that the SiO_2 values of the andesite analyses are normally distributed, in which case the normal distribution curve would then be used to approximate the actual distribution shown in Fig. 1.1.

The shapes of frequency distributions can vary considerable. For example, they may be symmetrical or asymmetrical about a central value. If asymmetrical they are either positively skewed when the distribution has a long tail to the right of the maximum value, or negatively skewed when the long tail of the distribution is to the left. They may also be unimodal, bimodal or even trimodal, according to their number of maxima. Distributions with more than one maxima are usually regarded by statisticians as representing samples drawn from more than one population.

1.3.1 *The normal distribution*

For many variables it is assumed that the population follows the normal distribution, which is symmetrical and determined by two parameters. The *mean*, μ (Greek letter mu), fixes the location of the centre of the distribution along the x axis, while the *standard deviation*, σ (Greek letter sigma), determines the spread of the curve and is, in fact, the scale of measurement along the x axis. The square of the standard deviation, σ^2, is known as the *variance*. If σ is small the curve is narrow and tall, if large the curve is broad and low. This is illustrated in Fig. 1.2, where three normal distributions with different means and standard deviation are shown.

The height of the normal distribution curve, or its density function, is given by the expression:-

$$y = \frac{1}{\sigma\sqrt{2\pi}} \exp\left[-(x - \mu)^2 / 2\sigma^2\right] \tag{1.2}$$

Integrating y over a particular range of x then gives the area under the curve and hence the probability of such a range of values of x occurring. Fortunately, this does not have to be done every time, as the areas are

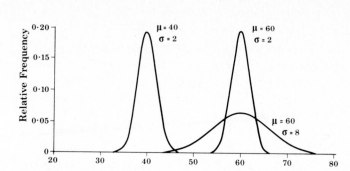

Fig. 1.2. Examples of three normal distribution curves. Note how the mean value μ controls the location of the curve, while the standard deviation σ controls its shape.

tabulated in most statistical textbooks. However, as there are an infinite number of possible normal distributions, with different values of μ and σ, the tables are for the specific normal distribution with μ = 0 and σ = 1, which is called the *standardized normal distribution*.

To use these tabulations for measurements taken from a population with mean μ and standard deviation σ, the values of x must first be scaled so that their mean is 0 and their standard deviation is 1, by using the relationship:-

$$z = (x - \mu) / \sigma \qquad (1.3)$$

where z is referred to as the *standard normal deviate*. Now although the tablulations are for the standardized normal distribution, the way in which the areas (probabilities) are tabulated are by no means standard, so that care must be taken when using tables to determine exactly which areas have been tabulated. For example, Dixon & Massey (1969) and Davis (1973) tabulate the areas from -infinity to z; Till (1974), Wonnacott & Wonnacott (1977) and Walpole & Myers (1978) tabulate areas from z to infinity or if z is negative from -z to -infinity; while Snedecor & Cochran (1967) tabulate areas from 0 to z. For conformity with other Tables in the Appendix, the areas tabulated in Table A2 are from z to infinity.

Finally, it is pertinent to ask why the normal distribution is so commonly used. The reasons are various but the most important are:-
1. Many natural measurements are approximately normally distributed (see Figs. 1.1 and 1.7).
2. Many that are not normal can be made normal by a simple transformation, such as taking logarithms or square roots (see also section 1.3.5).
3. Many quantities of interest are linear functions of variables and even if the variables are not normally distributed, the functions will tend to be normally distributed. Among such functions are many derived statistics and petrological values, such as means, normative and cation data.
4. The normal distribution is relatively easy to manipulate mathematically.

Numerical example

To illustrate the use of Table A2, assume that SiO_2 in a population of andesites is normally distributed with a mean of 58.86 and standard deviation of 4.28. What percentage of the population would be expected to have SiO_2 greater than 63? Using equation 1.3 we convert 63 to a standard normal deviate:-

$z = (63.0 - 58.86) / 4.28 = 0.97$

For z = 0.97, Table A2 gives a value of 0.1660, which means that 16.60% of the population would be expected to have SiO_2 greater than 63, i.e the rocks could be dacites. This is illustrated in Fig. 1.3. Similarly, if we wish to know the percentage of the population with SiO_2 less than 56, we find the new value of z:-

$z = (56.0 - 58.86) / 4.28 = -0.67$

As the normal distribution is symmetrical, we ignore the sign when using Table A2 which gives us an area of 0.2514 to the left of -z, so that 25.14% is the required answer. This limit is also shown in Fig. 1.3. The percentage of the andesite population with SiO_2 between 56 and 63, i.e. the basaltic andesites and andesites, would then be given by:-

$100.0 - 25.14 - 16.60 = 58.26$

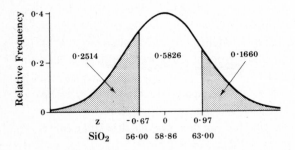

Fig. 1.3. Example of one use of the normal distribution curve. The curve is for a hypothetical population of andesites in which SiO_2 has a mean of 58.86% and a standard deviation of 4.28. The right-hand shaded area of 0.1660 (from Table A2 using z = 0.97) gives the probability of finding andesites with SiO_2 values greater than 63%. The left-hand shaded area of 0.2514 (from Table A2 using z = 0.67) gives the probability of finding andesites with SiO_2 values less than 56%.

1.3.2 *The lognormal distribution*

This distribution tends to occur with variables that have a natural lower limit, such as zero, and is related to the normal distribution in a simple way. If the variable x is lognormally distributed, $\log_e x$ is normally distributed.

The height of the lognormal distribution is given by the expression:-

$$y = \frac{1}{x\,\sigma_l\,\sqrt{2\pi}}\exp\left[-(\log_e x - \mu_l)^2\,/\,2\sigma_l^{\;2}\right] \qquad (1.4)$$

where μ_l is the mean value of the natural logarithms of the x's and σ_l is their standard deviation. Plotted on a linear scale this lognormal function is always positively skewed as shown in Fig. 1.4. The amount of skew is controlled by both μ_l and σ_l; the smaller the value of σ_l, the smaller the amount of skew and the more normally distributed the distribution becomes. Note, however, that unlike the normal distribution, a change in either the mean or standard deviation causes a change in both the location and shape of the distribution.

If x is lognormally distributed, the variable w – x has a negatively skewed lognormal distribution, with a natural upper limit of w as shown in Fig. 1.4. Some negatively skewed distributions of petrological variables are shown in Fig. 1.7, although Chayes (1976), feels that they are rather uncommon. Not all of these distributions are lognormal.

When a lognormal distribution is encountered or suspected, logarithms of all the values are often taken and the distribution of these transformed variables regarded as being normal. However, as Koch & Link (1970) point out, in some cases this is not always the best strategy to adopt.

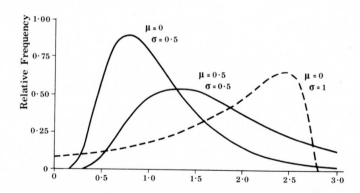

Fig. 1.4. Some examples of lognormal distributions (solid curves). Note how a change in either the mean μ, or the standard deviation σ, changes both the shape and location of the curve. The dashed curve is for a negative lognormal distribution.

1.3.3 *The binomial distribution*

This is a discrete distribution in which the value of the variable is limited to positive integers and is applicable to a particular type of sampling procedure, known as a binomial experiment, which must satisfy the following conditions:-
1. The experiment consists of n repeated trials.
2. Each trial may have one of two outcomes, usually referred to as success or failure e.g. quartz or not quartz, brown or white.
3. The probability, p, of a success is constant throughout the experiment. This means that if objects are selected in each trial they must be replaced before the next selection is made. If not the experiment is said to be without replacement, and the *hypergeometric distribution* should be used (Walpole & Myers, 1978).
4. The outcome of one trial does not depend upon the outcome of previous trials.
The probability that there will be x successes in n trials is then given by the expression:-

$$P(x) = \frac{n!}{x!(n-x)!}\, p^x\, (1-p)^{n-x} \tag{1.5}$$

If both np and n(1 - p) are greater than 5, the normal distribution, with a mean of np and a variance of np(1 - p), may be used to approximate the binomial distribution (Walpole & Myers, 1978, p.126). Note that in this equation either x or n - x can be zero, and by definition that 0! = 1.

In this form the binomial distribution has little application in petrology although Chayes (1956) has applied it to modal analysis.

Numerical example

The CLAIR data file of 25924 analyses contains 3594 analyses of rocks called basalt (Le Maitre, 1976a), so that the probability of selecting a single basalt analyses from the file is 3594/25924 or approximately 0.139. The probability of drawing a random sample of 7 analyses from the file (n = 7) in which 2 of the analyses will be of basalts (x = 2) is then given by equation 1.5 as:-

$$P(2) = \frac{7!}{2!5!} \times 0.139^2 \times 0.861^5 = 0.192$$

which is approximately 1 chance in 5. Similarly, the probability of only finding 1 basalt analysis in a sample of the same size is 0.396. If required, the probability of finding 2 or less basalt analyses in a sample of size 7 can be found by adding together the probabilities for x = 0, 1 & 2. The reader may care to verify that this comes to 0.939, or over 9 chances in 10. This distribution is illustrated in Fig. 1.5. Many textbooks tabulate these binomial

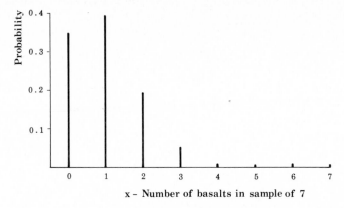

x – Number of basalts in sample of 7

Fig. 1.5. An example of the binomial distribution to show the probability of finding a specified number of basalt analyses in a sample of size 7 drawn from the CLAIR data file. The probability of selecting one basalt (a success) is 0.139.

probabilities for selected values of p and n in various ways, e.g. Dixon & Massey (1969, p.567) give individual probabilities, while Walpole & Myers (1978, p.509) give cumulative values.

1.3.4 The Poisson distribution

This is also a discrete distribution and may be derived as a limiting case of the binomial distribution (Walpole & Myers, 1978, p.102) or by completely independent methods. It is applicable to situations where the object of interest is the number of events, such as x-ray counts or volcanic eruptions, occurring in a specified region or time interval and must satisfy the following conditions:-

1. The number of events in one region or time interval is independent of those occurring in any others.
2. The probability of a single event occurring within an extremely small region or time interval is proportional to the size of the region or length of the time interval.
3. There is a negligible probability of more than one event occurring in the extremely small region or short time interval.

In practice the specified time interval can be of any size from microseconds to millions of years. Similarly, the regions may be lengths, areas, volumes or even n-dimensional hyperplanes. The probability that there will be x events in the specified region or time interval is then given by:-

$$P(x) = (e^{-\mu} \mu^x) / x! \qquad\qquad (1.6)$$

where μ is the mean number of events per region or time interval and e is the constant 2.71828... This distribution has the interesting property that its mean and variance are equal. As x becomes large the distribution also approximates the normal distribution. Further information on both this and the binomial distribution may be obtained from any of the standard texts e.g. Dixon & Massey (1969), Snedecor & Cochran (1971), Walpole & Myers (1978).

As will be seen in Chapter 11, the main use of the Poisson distribution in petrology is its application to the errors involved in x-ray counting techniques.

Numerical example

The background count for an x-ray counter is 2 counts per second. What is the probability of obtaining no counts in one second? Using equation 1.6 we have:-

$$P(0) = e^{-2}2^0 / 0! = 0.135$$

The reader may care to verify that $P(2) = 0.271$ and $P(4) = 0.090$. This distribution is illustrated in Fig. 1.6. Tabulations, some of which are cumulative, are also available for this distribution in many textbooks, e.g. Dixon & Massey (1967, p.533), Walpole & Myers (1978, p.510-512).

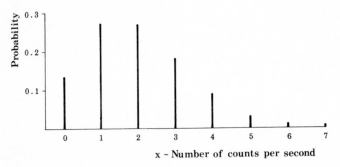

Fig. 1.6. An example of the Poisson distribution to determine the probability of obtaining a specified number of x-ray counts per second, when the mean rate is 2 counts per second.

1.3.5 Transformation of variables

One of the reasons for transforming variables is to make the actual frequency distribution more like the normal distribution, so that the statistical tests which are based on the normal distribution can be applied. In the case of a bimodal or multimodal distribution, no transformation will make the distribution approximately normal, but with skewed unimodal distributions there are several transformations that can be tried. Emphasis must be placed on the word tried, as a transformation is largely an empirical approach - if it works, use it. Another reason for transforming variables is to stabilize the variances, as often done in the analysis of variance (Chapter 3).

Many specific transformations have been suggested. One is the *log transformation,* either as $z = \log(x)$ or $z = \log(x + w)$, where w is a constant. This can make a positively skewed unimodal distribution (x) much more normal, but the distribution of z may still be positively or negatively skewed. To correct for a positive skew in z, w should be negative and vice versa. For very positively skewed distributions, it is possible for x + w to be negative and hence z will not exist. In such a case the distribution is said to be *censored,* as certain values can not be represented. Distributions can also be censored if values fall below the lower limit of detection. Methods of dealing with such distributions can be found in Miesch (1967b).

Other transformations that can be tried are the *square root transformation,* for data believed to have a Poisson distribution. This can take the form $z = \sqrt{(x + w)}$ or, if the values of x are small, $z = \sqrt{x} + \sqrt{(x + w)}$, where w is a constant. A transformation that can be used for proportions or percentages, is the *angular* or *arcsin transformation,* which was developed for the binomial distribution, and takes the form $z = \arcsin(x)$, thus spreading out the values near 0 and 1. More information concerning these methods can be found in Miller & Kahn (1962), Snedecor & Cochran (1967) and Dixon & Massey (1969).

As an alternative, Howarth & Earle (1979) advocate the use of one of the generalized power transformations of Box & Cox (1964), particularly for the preprocessing of geochemical data. The recommended form is:-

$$z = (x^\lambda - 1) / \lambda \tag{1.7}$$

where the choice of λ (Greek letter lambda) can be made to optimize the shape of the distribution of the transformed variable. Howarth & Earle (1979) also include a Fortran program which will do the optimization in a variety of ways. For particular values of λ the transformation is equivalent to some specific transformations e.g. when $\lambda = 0.5$ it is the square root transformation and when $\lambda = -1$ it is the reciprocal transformation. In the limiting case when λ goes to zero and z appears to be indeterminate, equation 1.7 can be shown to be equivalent to the natural log transformation.

1.3.6 Some observed petrological frequency distributions

Some examples of frequency distributions encountered with petrological data are shown in histogram form in Fig. 1.7. These were selected from histograms produced for 49 normative values, oxides and ratios from the following 4 sets of data:- 38098 complete analyses in the CLAIR and PETROS data files; 2485 granites in the CLAIR data file; 2600 andesites in the CLAIR data file; and 3594 basalts in the CLAIR data file. The latter three groups are the data used in Le Maitre (1976a).

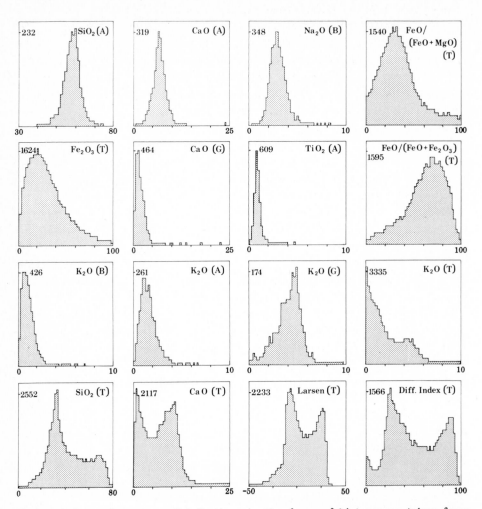

Fig. 1.7. Some frequency distributions in the form of histograms taken from data in the CLAIR and PETROS data files. The number in the top left-hand corner of each histogram is a frequency, while the numbers along the base give the scale of the variable. The abbreviations in brackets are as follows:- T = 38098 complete analyses from the CLAIR and PETROS data files; B = 3594 basalts from CLAIR; A = 2578 andesites from CLAIR; G = 2485 granites from CLAIR. The distributions in the top row are approximately symmetrical, although $100FeO/(FeO+MgO)$ is truncated; in the 2nd row they are all skewed; the 3rd row illustrates how the distribution of one oxide may vary in different rock types; and the bottom row shows 3 bimodal and 1 trimodal distribution.

From these data and experience with many others, the following generalizations can be made:-

1. Many petrological variables are symmetrically distributed with only minor amounts of skew, particularly if single rock types are examined, rather than rock suites. Consistent examples are Al_2O_3, Na_2O, Ab (directly related to

Na_2O) and the $100FeO/(FeO+MgO)$ ratio in the normative components, although this distribution is often truncated at 0, 100 or both. Several of these distributions are also approximately normal.

2. Major amounts of skew occur with those variables whose mean and standard deviation are similar in value e.g. TiO_2, P_2O_5, MnO and H_2O+. Included in this category are most of the trace elements and those elements which Ahrens (1954) studied in formulating his fundamental law of geochemistry that "the concentration of an element is lognormally distributed in a specific igneous rock". Note that the $100FeO/(FeO+Fe_2O_3)$ ratio exhibits a consistently negative skew, even though its standard deviation is usually considerably smaller than its mean. This no doubt reflects the fact that many igneous rocks are oxidized, thus enhancing the low values of the $100FeO/(FeO+Fe_2O_3)$ ratio.

3. Bimodal and even trimodal distributions are usually confined to data drawn from rock series. Some variables that are bimodally distributed when a wide variety of igneous rocks are considered together, are SiO_2, CaO, normative Q and An, the Larsen function and to a lesser extent Na_2O and K_2O. Both the differentiation index and normative colour index are weakly trimodal.

4. Variables do not always have the same distribution. For example, K_2O has a negative skew in granites; a weak positive skew in andesites; a strong positive skew in basalts; and is bimodal in igneous rocks as a whole.

The frequency distribution of values determined on rock standards (Steele & Hansen, 1979) tend to be normal or lognormal with few of the irregularities found in petrographically defined groups of rocks. This is to be expected as the main sources of error are in splitting the powder and in the analytical procedures.

CHAPTER 2

ESTIMATING AND TESTING POPULATION PARAMETERS

2.1 INTRODUCTION

In the previous chapter it was assumed that the population parameters μ and σ were known, but this is rarely the case. Instead we have to make do with estimates of the population parameters derived from values calculated from actual samples. These values are called *statistics* and by convention are usually denoted by Latin letters e.g. $\bar{x}$ and s for the sample mean and standard deviation, respectively. Because many samples can be drawn from any population, the statistics will vary from sample to sample and will, therefore, have their own distributions. Some of the results most likely to be of use are given below. For the derivation of the expressions, many excellent statistical texts are available e.g. Dixon & Massey (1969), Snedecor & Cochran (1967), Walpole & Myers (1978), Wonnacott & Wonnacott (1977), to name but a few.

2.2 ESTIMATION OF POPULATION MEAN μ, AND STANDARD DEVIATION σ

If we have a sample of n values x_1, x_2, ... x_n, drawn at random from a normally distributed population, the unbiased estimate of μ, the population mean, is given by $\bar{x}$ (called x-bar), the sample mean, which is defined as:-

$$\bar{x} = (x_1 + x_2 + ... + x_n) / n = \sum x_i / n \qquad (2.1)$$

where $\sum x_i$ (called sigma x_i) stands for the sum of all values of x_i.

Similarly, the unbiased estimate of the population variance, σ^2, is given by the sample variance s^2, which is defined as:-

$$s^2 = \sum (x_i - \bar{x})^2 / (n - 1) \qquad (2.2)$$

This expression is the sum of squares of the deviations of each observation from the sample mean divided, not by n as in equation 2.1, but by (n - 1). This factor is known as the *degrees of freedom* (abbreviated d.f.), and is to correct for bias that would occur in the estimate, if n was the divisor (Walpole & Myers, 1978, p.191). Degrees of freedom are used in calculating many statistics and are somewhat analogous to the degrees of freedom in the chemical phase rule. They can be explained in the following way; n observations have n d.f. as the observations can all be chosen independently, but calculating $\bar{x}$ in equation 2.2

uses up 1 d.f. so that only n - 1 remain for the sum of squares of deviations from $\bar{x}$. In general, the number of d.f. is given by the number of observations minus the number of estimates made from the observations that are necessary to calculate the statistic.

The *sum of squares* of deviations from the mean is used in many statistical expressions and will be abbreviated SS, with a subscript to indicate the variable involved. However, equation 2.2 is not very practical to use as it involves calculating $\bar{x}$ before the individual deviations can be calculated, which means handling the data twice. A more convenient form of equation 2.2, and the one used by electronic calculators, can be obtained by the following manipulation:-

$$
\begin{aligned}
SS_x &= \sum (x_i - \bar{x})^2 \\
&= \sum (x_i^2 - 2x_i\bar{x} + \bar{x}^2) \\
&= \sum x_i^2 - 2\bar{x}\sum x_i + n\bar{x}^2 \\
&= \sum x_i^2 - 2(\sum x_i)(\sum x_i) / n + n(\sum x_i / n)^2 \quad \text{using equation 2.1} \\
&= \sum x_i^2 - (\sum x_i)^2 / n
\end{aligned}
$$

(2.3)

which then gives the variance as:-

$$
s^2 = \left[\sum x_i^2 - (\sum x_i)^2 / n \right] / (n - 1)
$$

(2.4)

which involves accumulating $\sum x_i$ and $\sum x_i^2$ at the same time. The standard deviation can then be obtained by taking the square root of the variance.

In many applications the absolute value of a standard deviation has little tangible meaning so that it is often divided by the mean to give what is known as the *relative standard deviation, relative error* or *coefficient of variation*, which may be quoted as a fraction or as a percentage, when it is also known as the *percentage relative error*. If the population parameters are not known the coefficient of variation may be estimated as:-

$$
E = s / \bar{x}
$$

(2.5)

Worked examples of these methods are given in Table 2.1, using a small sample of SiO_2 values drawn at random from the group of 2578 andesites in the CLAIR data file (Le Maitre, 1976a).

2.3 STANDARD DEVIATION OF THE SAMPLE MEAN

If we draw repeated random samples of size n from a population, the values of $\bar{x}$ will have a standard deviation of $s_{\bar{x}}$, which is called the *standard error*, and is given by:-

$$
s_{\bar{x}} = s / \sqrt{n}
$$

(2.6)

If n is greater than 30, the distribution of $\bar{x}$ will be near enough to normal whatever the distribution of the population; if not, the distribution of $\bar{x}$ is only normal if the population is nearly normal (Walpole & Myers, 1978, p.167).

TABLE 2.1

Calculation of sample mean and standard deviation using the SiO_2 values of a random sample of 7 andesite analyses taken from 2578 andesite analyses in the CLAIR data file.

Specimen	x	$x - \bar{x}$	$(x - \bar{x})^2$	x^2
1	59.51	4.54	20.6116	3541.4401
2	58.35	3.38	11.4244	3404.7225
3	56.39	1.42	2.0164	3179.8321
4	56.25	1.28	1.6384	3164.0625
5	54.90	-0.07	0.0049	3014.0100
6	49.21	-5.76	33.1776	2421.6241
7	50.18	-4.79	22.9441	2518.0324
Totals	384.79	0.00	91.8174	21243.7237

Mean $\quad\quad\quad \bar{x} = 384.79 / 7 = 54.97 \quad$ (using equation 2.1)

Variance $\quad\; s^2 = 91.8174 / 6 = 15.3029 \quad$ (using equation 2.2)

S.D. $\quad\quad\quad s = 3.91$

Variance $\quad\; s^2 = (21243.7237 - 384.79^2 / 7) / 6 \quad$ (using equation 2.4)

$\quad\quad\quad\quad\quad\; = 91.8174 / 6 = 15.3029$

S.D. $\quad\quad\quad s = 3.91$

Relative error $\quad E = 3.91 / 54.97 = 0.071$ or 7.1% $\quad$ (using equation 2.5)

2.4 PRINCIPLES OF HYPOTHESIS TESTING

Walpole & Myers (1978, p.238) define a statistical hypothesis as "an assumption or statement, which may or may not be true, concerning one or more populations". Although we can never be completely certain of the validity of a hypothesis, unless we are able to examine the entire population, we can make statements about the probability of it being true or false, based on statistics calculated from samples taken at random from the population. However, if the *sample is biased*, then the *statements will be applicable to a new sub-population* from which the sample could have been randomly selected.

If the statistics conflict with the hypothesis we reject it, but if they do not conflict we accept it. However, it is important to realize that acceptance of an hypothesis merely indicates that we have no evidence to reject it, which is not the same as proving that the hypothesis is true.

In practice we set up a *null hypothesis*, denoted H_0, which we hope to reject, and test this against an *alternate hypothesis*, denoted H_1. If H_0 is rejected, we accept H_1. For example, the null hypothesis may be that the population mean μ_1 is equal to a specified value μ which is written as:-

$$H_0 : \mu_1 = \mu \quad\quad\quad\quad\quad\quad\quad\quad\quad\quad\quad\quad\quad\quad\quad (2.7)$$

Before this can be tested we must specify the alternate hypothesis which could be that the population mean μ_1 is not equal to μ which is written as:-

$$H_1 : \mu_1 \neq \mu \qquad \qquad (2.8)$$

or that the population mean μ_1 is less than μ or:-

$$H_1 : \mu_1 < \mu \qquad \qquad (2.9)$$

If the null hypothesis is rejected when it is true we have committed a *type I error*, but if we accept it when it is false we have committed a *type II error*. Both of these errors are related and in general if the probability of one decreases the probability of the other increases (see Dixon & Massey, 1969, p.83-89 for an excellent discussion). The probability of committing type I errors is denoted by α, (Greek letter alpha) and is called the *level of significance*. This must also be specified before any testing can take place. A test is usually said to be *significant* if H_0 is rejected at the 0.05 level of significance, and *highly significant* if rejected at the 0.01 level.

Let us consider the null hypothesis that $H_0 : \mu_1 = \mu$ against the alternative hypothesis $H_1 : \mu_1 < \mu$. This entails setting up a *one-tailed test* with a critical region entirely to one-side of the population distribution as shown in Fig. 2.1a. If the value of the sample mean is to the right of the critical region, which is the shaded portion of the distribution with an area of α, H_0 is accepted, as there is no evidence to suggest that μ_1 is not equal to μ. However, if the value of the sample mean is in the critical region, H_0 is rejected as the probability of μ being so far from μ_1, quite by chance, is beyond our specified level of significance. In this instance, H_1 would then be accepted.

If we had chosen the alternative hypothesis $H_1 : \mu_1 \neq \mu$, which provides two possibilities that $\mu_1 < \mu$ or $\mu_1 > \mu$, a *two-tailed test* would be used with two critical regions as shown in Fig. 2.1b. Note that in this case the area of each critical region is $\alpha/2$. If the value of the sample mean lies between the

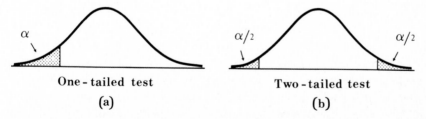

One-tailed test   **Two-tailed test**

(a)   (b)

Fig. 2.1. Diagram to illustrate the different critical regions (the shaded areas) in (a) a one-tailed and (b) a two-tailed test at a level of significance of α. H_0 is rejected if the test value falls in a critical region, otherwise it is accepted.

two critical regions, H_0 is accepted, otherwise it is rejected. The two-tailed test is also the basis for setting up *confidence limits*, which are two values between which a particular population parameter has a certain probability of occurring.

Finally, it is important to understand the full implications of a statement that a particular test is *"statistically significant"*. These are that one is fairly sure that the *null hypothesis is false*, but if it is *actually true*, then either *a rare event must have occurred* or *the sample is biased.*

2.5 CONFIDENCE LIMITS FOR POPULATION MEAN WHEN σ KNOWN

Fig. 2.2 shows a normal distribution curve with **mean** μ and standard deviation $\sigma/\sqrt{n}$ i.e. the sample distribution of $\bar{x}$. Using Table A2 in the Appendix, values of $z_{\alpha/2}$ can be obtained which give an **area** of $\alpha/2$ under the normal curve, so that $100(1 - \alpha)\%$ of the area lies between the two critical regions which are shaded. If a sample of size n is taken from the population, there is a $100(1 - \alpha)\%$ probability that the value of $\bar{x}$ will lie between the two critical regions. Using equation 1.2 this can be written as **two inequalities:-**

$$\bar{x} - z_{\alpha/2}\, \sigma/\sqrt{n} < \mu < \bar{x} + z_{\alpha/2}\, \sigma/\sqrt{n} \tag{2.10}$$

The value $100(1 - \alpha)\%$ is known as the *confidence level,* while the two values either side of the inequality are called the *confidence limits.* When they are symmetrical about a mean value, they are often expressed in the alternate form as:-

$$\bar{x} \pm z_{\alpha/2}\, \sigma/\sqrt{n} \tag{2.11}$$

Numerical example

Using the mean sample value of 54.97 for the SiO_2 data in Table 2.1 and still assuming that the population standard deviation is 4.28, we will calculate the 95% confidence limits for the population mean μ. From Table A2, $z_{0.025} =$ 1.960 so that we may write the confidence limits as:-

$54.97 \pm (1.960 \text{ x } 4.28) / \sqrt{7} = 54.97 \pm 3.17$
or
$51.80 < \mu < 58.14$

We can then state that at the 95% confidence level the value of the population mean will be between 51.80 and 58.14. In fact we know μ to be 58.86, which is just outside the range, so that this particular example is one of the 5 expected in every 100 samples of size 7 in which the population value will be outside the range of the confidence limits. However, if we repeat the calculation at the 99% confidence level, we find that $z_{0.005} = 2.576$, so that the confidence limits are then 50.82 and 59.12, which does include the population mean.

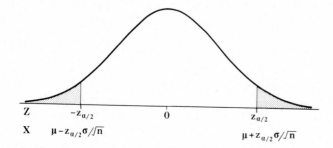

$$
\begin{array}{cccc}
Z & -z_{\alpha/2} & 0 & z_{\alpha/2} \\
X & \mu - z_{\alpha/2}\sigma/\sqrt{n} & & \mu + z_{\alpha/2}\sigma/\sqrt{n}
\end{array}
$$

Fig. 2.2. Diagram to illustrate the meaning of confidence limits for the population mean. The normal curve has a mean of μ and standard deviation of $\sigma/\sqrt{n}$, i.e. it is the sampling distribtuion of x. $100(1 - \alpha)\%$ of samples taken of size n would be expected to give values of x that would lie between the two critical regions.

2.6 SAMPLE SIZE

A frequent problem is what size of sample should be taken to be confident, at a specified level of probabilty, that the sample mean $\bar{x}$, will differ from the population mean μ, by less than a specified amount e. Using equation 2.10 this is equivalent to stating that $z_{\alpha/2}\sigma/\sqrt{n} = e$ or:-

$$n = (z_{\alpha/2}\sigma / e)^2 \tag{2.12}$$

Again this assumes that σ is known, but if σ has been estimated using s, the results will be reasonable as long as the sample size from which s was estimated is not too small (Walpole & Myers, 1978, p.197).

Numerical example

How many randomly chosen andesites would have to be selected to be able to state that the sample mean of SiO_2 differs from the population mean by less than 0.5 at the 95% confidence level? Using equation 2.12 we get:-

$$n = (1.96 \times 4.28 / 0.5)^2 = 282$$

Check that at the 99% confidence level, we would have to select 483 analyses.

2.7 CONFIDENCE LIMITS FOR POPULATION MEAN WHEN σ UNKNOWN

In all the calculations so far, it has been assumed that the population standard deviation σ is known, but this, of course, is rarely true. In practice we have to make do with the sample standard deviation s, based on $(n - 1)$ degrees of freedom, as an estimate of σ, which introduces an added uncertainty into our calculations. This problem is overcome by the *Student's t distribution* where t is defined as:-

$$t = \frac{\bar{x} - \mu}{s / \sqrt{n}} \tag{2.13}$$

Although t cannot be calculated unless an assumption is made about the value of μ, the sampling distribution of t can be calculated. Like the normal distribution, it is also symmetrical. The areas under one tail of the t distribution, or the critical values, are given in Table A3 in the Appendix, for a range of probabilities and degrees of freedom. Note that as the number of d.f. increases the value of t_α approaches the value of z_α until when the number of d.f. is infinite, $t_\alpha = z_\alpha$. This provides a convenient method of obtaining z values directly for specific probabilities by using Table A3 with infinite d.f. instead of by interpolation in Table A2. The $100(1 - \alpha)\%$ confidence limits on μ can now be stated by rewriting equation 2.13 as two inequalities:-

$$\bar{x} - t_{\alpha/2} s / \sqrt{n} < \mu < \bar{x} + t_{\alpha/2} s / \sqrt{n} \tag{2.14}$$

which has the same form as equation 2.10, with $z_{\alpha/2} \sigma$ replaced by $t_{\alpha/2}$ s. Alternatively, we may express this as:-

$$\bar{x} \pm t_{\alpha/2} s / \sqrt{n} \tag{2.15}$$

Numerical example

We will now determine the 95% confidence limits on the population mean of SiO_2, using the sample mean of 58.86 and the sample standard deviation of 3.91. Entering Table A3 with 6 d.f. we find $t_{0.025} = 2.447$, so that we may write the confidence limits as:-

$54.97 \pm (2.447 \times 3.91) / \sqrt{7} = 54.97 \pm 3.62$

or

$51.35 < \mu < 58.59$

As can be seen these confidence limits are wider than those given for the 95% confidence level in section 2.5, due to the added uncertainty of the correct value of the population standard deviation.

2.8 TESTING FOR A DIFFERENCE BETWEEN TWO MEANS

Student's t can also be used to test if there is a difference between two population means. If two samples of size n_1 and n_2, have means $\bar{x}_1$ and $\bar{x}_2$, and variances s_1^2 and s_2^2, a modified form of equation 2.10 can be written in which:-

$$t = \frac{\bar{x}_1 - \bar{x}_2}{s_p \sqrt{(1/n_1 + 1/n_2)}} \tag{2.16}$$

where s_p is called the pooled variance and is given by:-

$$s_p = \frac{(n_1 - 1)s_1{}^2 + (n_2 - 1)s_2{}^2}{n_1 + n_2 - 2} \tag{2.17}$$

It is assumed that the two population variances are equal, and this can be tested as shown in the next section. If they are not equal, slightly different equations should be used (e.g. see Walpole & Myers, 1978, p.205).

The test can be represented formally as $H_0 : \mu_1 = \mu_2$ with a choice of alternate hypotheses e.g. $H_1 : \mu_1 \neq \mu_2$, $H_1 : \mu_1 > \mu_2$, etc.. When the value of t, which has $(n_1 + n_2 - 2)$ d.f., has been calculated, Table A3 can be used to see whether or not the value exceeds that for the desired level of significance and the chosen alternate hypothesis.

Numerical example

In a different sample of 9 andesite analyses, SiO_2 has a mean of 60.53 and standard deviation of 5.79. Does the population mean from which this sample was drawn differ significantly from the population mean of the previous sample? The pooled variance is given by equation 2.17:-

$$s_p = (6 \times 3.91^2 + 8 \times 5.79^2) / (7 + 9 - 2) = 25.71$$

and using equation 2.16 t is:-

$$t = (60.53 - 54.97) / \left[25.71 \sqrt{(1/9 + 1/7)} \right] = 0.43$$

At the 95% confidence level Table A3 gives the critical value of $t_{0.025}$ with 14 d.f. as 2.145. As this greatly exceeds the calculated value of 0.43, the evidence does not support the conclusion that the two population means are different from each other.

2.9 VARIANCE RATIO TEST AND F DISTRIBUTION

Two variances can be tested for equality by forming the statistic F as follows:-

$$F = s_1{}^2 / s_2{}^2 \tag{2.18}$$

where $s_1{}^2$ and $s_2{}^2$ are the variances of two independent samples of size n_1 and n_2, supposed to have been drawn from the same normal population. The null hypothesis to be tested is $H_0 : s_1{}^2 = s_2{}^2$ against the alternative hypothesis $H_1 : s_1{}^2 > s_2{}^2$. The distribution of F, obtained by theoretically taking all possible samples of size n_1 and n_2 from the population, is known and the critical values are given in Table A4 in the Appendix. Note that this table must be entered with two d.f.; $(n_1 - 1)$ for the numerator, and $(n_2 - 1)$ for the denominator.

Numerical example

One sample of 9 andesite analyses has a standard deviation of 5.91, and another of 7 has a standard deviation of 3.00. At the 95% confidence level, could the variances of the two populations from which these samples were drawn be consider to be the equal? Using equation 2.18 we have:-

$$F = 5.91^2 / 3.00^2 = 3.88$$

with 8 d.f. for the numerator and 6 d.f. for the denominator. Table A4 gives the critical value of $F_{0.05} = 4.15$. As this is in excess of 3.88 the evidence does not support the conclusion that the two variances are different.

2.10 GOODNESS-OF-FIT OR CHI-SQUARE TEST

This is an extremely useful test and has many uses. Basically it tests how well an observed set of frequency data fits an expected set of frequency data. This is done by computing the statistic χ^2 (Greek letter chi) as:-

$$\chi^2 = \sum_{i=1}^{k} \frac{(O_i - E_i)^2}{E_i} \qquad (2.19)$$

where O_i and E_i are the observed and expected frequencies in the ith arbitrarily chosen class, respectively, and k is the number of classes. The value of χ^2 can then be tested against the critical values for chi-square given in Table A5 in the Appendix. Alternatively, it is often more convenient to calculate the probability of obtaining a particular value of χ^2 directly by using simple algorithms available with many pocket calculators and in many computer packages e.g. subroutine MDCH in IMSL (International Mathematical and Statistical Libraries Inc.). The d.f. associated with χ^2 is slightly more complicated than usual and is given by:-

$$d.f. = k - q \qquad (2.20)$$

where q is the number of quantities obtained from the observed data that are used in calculating the expected frequencies. The example will help to clarify the meaning of q.

Numerical example

The SiO_2 values of a sample of 49 andesites taken at random from the CLAIR data file are given in Table 2.2. Could these values have been obtained from a normally distributed population with mean 58.86 and standard deviation 4.28? To test this hypothesis the data must be divided into any convenient number of classes. Let us choose 8 and divide the normal distribution into 8

TABLE 2.2

Use of goodness-of-fit or chi-square test to see if the SiO_2 values of 49 randomly chosen andesites from the CLAIR data file could have come from a normally distributed population with a mean of 58.86 and standard deviation of 4.28. The boundary values of the eight classes are chosen so that equal numbers of observations should occur in each class if the data have been randomly sampled from a normally distributed population, i.e. the expected value in each class (Exp.) is 49/8 = 6.125.

SiO_2 range of classes	SiO_2 values of individual andesites			Observed number	$\dfrac{(Obs. - Exp.)^2}{Exp.}$
53.936	52.93 49.21 51.33	53.22 50.18	52.61 53.56	7	0.125
53.936 – 55.973	55.28 55.18 54.76	55.74 54.90 54.76	54.60 54.83 54.24	9	1.349
55.973 – 57.496	56.99 56.29 56.45	56.39 56.82	56.25 57.35	7	0.125
57.496 – 58.860	57.75 58.16 58.35	57.53 57.68	58.00 57.66	7	0.125
58.860 – 60.224	60.22 59.97	59.68 59.51	59.80 60.05	6	0.003
60.224 – 61.747	60.62 61.50	61.68 61.65	60.83 61.63	6	0.003
61.747 – 63.784	63.61 61.81	62.24	62.82	4	0.737
63.784	72.35	69.15	67.60	3	1.594

$$\chi^2 = 4.061$$

portions of equal area. This is done by finding the values of z from Table A2, for which the areas under the normal curve are 0.125, 0.25 and 0.375. By interpolation, the required values are 1.1505, 0.6745 and 0.3187, respectively. At this stage either all the SiO_2 values must be converted to standard normal deviates by using equation 1.3, or the values of z must be converted to SiO_2 values. Using the latter, as it involves less calculation, the z values convert to SiO_2 values of 4.924, 2.887 and 1.364, respectively. Adding and subtracting these values to the mean SiO_2 value of 58.86 we can then obtain the 6 values which, together with the mean value, define the boundaries between the 8 classes. These values are shown in Table 2.2 together with the SiO_2 values of the individual andesites and the observed numbers in each class. The expected number of andesites in each class is, of course, 49/8 or 6.125. Table 2.2 also gives the contribution that each class makes to the value of χ^2 which is 4.061. In this example the d.f. are 8 - 3 = 5, as the number of classes is 8 and the number of quantities obtained from the observed data that are used to calculate the expected frequencies is 3 (the total number of observations, the mean and standard deviation). At the 95% confidence level the critical value of χ^2 with 5 d.f. is 11.07 (Table A5), which greatly exceeds the calculated value of 4.061. We, therefore, conclude that there is no evidence to suggest that the data did not come from a normally distributed population.

Using an appropriate calculator, the probablility of obtaining a value of χ^2 greater than 4.061 with 5 d.f. can be calculated to be 54.1%, i.e. there is a slightly better than even chance.

CHAPTER 3

ANALYSIS OF VARIANCE - TESTING DIFFERENCES AMONG SEVERAL

MEANS

3.1 INTRODUCTION

In the previous chapter, methods of testing for differences between the means of two populations were given. However, it is often useful to be able to test for differences between the means of more than two populations simultaneously. For example, in studies involving measurements on standards, it may be desired to compare the results made by different analytical methods or laboratories. One method of doing this is known as *analysis of variance* as it splits the total sum of squares of the observations into several portions to test specific desired conclusions. The design of analysis of variance methods can be extremely flexible, and only some of the simpler methods immediately pertinent to petrology will be dealt with here. Readers interested in pursuing the topic further may refer in the first instance to texts such as Dixon & Massey (1969), Snedecor & Cochran (1971) or Walpole & Myers (1972).

3.2 ONE-WAY CLASSIFICATION

This is the simplest method and is an extension of the method of testing for differences between means given in section 2.8. In this design each observation is assumed to come from one of k populations (often called categories) with means μ_1, μ_2, ... ,μ_k and equal variances σ^2. The usual null hypothesis is that the means of all the categories are equal, but tests can also be made about specific means and functions of means (e.g. Dixon & Massey, 1969, p.162-167).

Basically the analysis of variance consists of estimating the population variance in two different ways and then comparing the results with a variance ratio or F-test as described in section 2.9. The reason for this approach is that if the means of the k categories are significantly different from each other, the two estimates of the population variance will also be significantly different. One method involves calculating the variance of the means of each category which, by using equation 2.6, also provides an estimate of the population variance. This estimate is called the *mean square between categories* and denoted by s_m^2. The other is a pooled estimate s_p^2, called the *within categories mean square,* and is based on an extension of equation 2.4. A third estimate can also be made by calculating the variance of all the observations as if they had come from one population. This is usually reported as a check, as this total sum of squares and d.f. are equal to the sum of the between and within sum of squares and d.f., respectively.

The generalized method of calculation is given in Table 3.1. If the categories come from populations with different means, s_m^2 will be considerably larger than s_p^2, so that the variance ratio used is always s_m^2/s_p^2. If this value exceeds the critical value of F for $k - 1$ and $N - k$ d.f. as given in Table A4 in the Appendix, the null hypothesis is rejected.

Numerical example

To illustrate this method some Sr values determined by 3 different analytical techniques (the categories) on the USGS standard W-1 have been taken from Flanagan (1976b, p.164-165) and are given in Table 3.2 together with details of the calculation and the analysis of variance table. The problem is to decide if there are any significant differences between the mean values of the 3 analytical methods at a confidence level of 95%. As can be seen the mean square between categories $s_m^2 = 2358.20$ and the within categories mean square $s_p^2 = 252.72$, which gives a value of F of 9.33 with 2 and 20 d.f. As this

TABLE 3.1

Generalized computational scheme for one-way analysis of variance, largely following the nomenclature of Dixon & Massey (1969, p.157). In the table below x_{ij} is the jth observation for the ith category. In this design the number of observations in each category can be different.

	Category					Total
	1	2	3	.	k	
	x_{11}	x_{21}	x_{31}	...	x_{k1}	
	x_{12}	x_{22}	x_{32}	...	x_{k2}	
	x_{13}	x_{23}	x_{33}	...	x_{k3}	
	...	...	...	...	...	
Column totals	T_{1+}	T_{2+}	T_{3+}	...	T_{k+}	T_{++}
Number Obs.	n_1	n_2	n_3	...	n_k	N

ONE-WAY ANALYSIS OF VARIANCE

Source of variation	Sum of Squares	Degrees of freedom	Variance or mean square	F-ratio
Means	$S_B = \sum (T_{i+}^2/n_i) - T_{++}^2/N$	$k - 1$	$s_m^2 = S_B/(k - 1)$	$F = s_m^2/s_p^2$
Within	$S_W = \sum\sum x_{ij}^2 - \sum (T_{i+}^2/n_i)$	$N - k$	$s_p^2 = S_W/(N - k)$	
Total	$S_T = \sum\sum x_{ij}^2 - T_{++}^2/N$	$N - 1$		

Note that: $S_T = S_B + S_W$

TABLE 3.2

Example of one-way analysis of variance using the values of Sr (ppm) determined by 3 different analytical methods on the USGS standard W-1 (Flanagan, 1976b). The null hypothesis is that the mean Sr values for the 3 analytical methods are equal.

	Optical spectrographic	Atomic absorption	X-ray fluorescence		Total
	186	235	189	209	
	182	178	180	200	
	170	215	192	180	
	156	202	190	196	
	125	205	183	184	
		195	190	189	
Totals	819	1230	2282		4331
Number	5	6	12		23
Mean	163.8	205.0	190.2		

$\sum (T_{i+}^2/n_i) = 819^2/5 + 1230^2/6 + 2282^2/12 = 820262.53$

$T_{++}^2/N = 4331^2/23 = 815546.13$

$\sum\sum x_{ij}^2 = 186^2 + 182^2 + .. +235^2 + 178^2 + .. + 189^2 + .. 189^2 = 825317.0$

$S_B = 820262.53 - 815546.13 = 4716.40 \qquad s_m^2 = 4716.40/2 = 2358.20$

$S_W = 825317.0 - 820262.53 = 5054.47 \qquad s_p^2 = 5054.47/20 = 252.72$

$S_T = 825317.0 - 815546.13 = 9770.87 \qquad F = 2358.20/252.72 = 9.33$

ONE-WAY ANALYSIS OF VARIANCE

Source of variation	Sum of Squares	Degrees of freedom	Variance or mean square	F-ratio
Means	4716.40	2	2358.20	9.33
Within	5054.47	20	252.72	
Total	9770.87	22		

Conclusion: F = 9.33 is significant as $F_{0.05} = 3.49$ with 2 and 20 d.f.

is considerably larger than the critical value of 3.49 obtained from Table A4, we must reject the null hypothesis that there are no differences between the means of the 3 sets of analyses. Even at a confidence level of 99% the null hypothesis would have been rejected as the test value of F is still greater than the critical value 5.85. Before the reader gets too despondent about the reliability of analytical results, it must be pointed out that a considerable amount of effort went into finding an example in which the means were significantly different.

3.3 TWO-WAY CLASSIFICATION WITH SINGLE OBSERVATIONS

This is a slightly more advanced design in which each observation belongs to two categories as set out in Table 3.3, and enables independent tests to be made concerning differences between the means for category 1 or 2. The assumptions made in this design are slightly different from those for the one-way classification. Although it is still assumed that the variance of each cell population is the same, it is also assumed that the mean of each cell is different and made up of components from three populations. One population has a constant mean for each cell while the other two, called the "column" and "row" effects, have constant means for each column and row, respectively. It is further assumed that the row and column effects do not interact with each other and are simply additive in their effects.

By analogy with the one-way classification, mean squares for categories 1 and 2 can be calculated as shown in Table 3.3 together with their respective sum of squares and d.f. These give measures of the variability of the row and

TABLE 3.3

Generalized computational scheme for two-way analysis of variance with single observations, largely following the nomenclature of Dixon & Massey (1969, p.168). In the table below x_{ij} is a single observation from the ith category 1 and the jth category 2.

Category 2	Category 1					Row Totals
	1	2	3	.	c	
1	x_{11}	x_{21}	x_{31}	...	x_{c1}	T_{+1}
2	x_{12}	x_{22}	x_{32}	...	x_{c2}	T_{+2}
3	x_{13}	x_{23}	x_{33}	...	x_{c3}	T_{+3}
.	...	...	...	...	...	...
r	x_{1r}	x_{2r}	x_{3r}	...	x_{cr}	T_{+r}
Column Totals	T_{1+}	T_{2+}	T_{3+}	...	T_{c+}	T_{++}

TWO-WAY ANALYSIS OF VARIANCE: SINGLE OBSERVATIONS

Source of variation	Sum of Squares	Degrees of freedom	Variance or mean square	F-ratio
Col means	$S_c = \sum (T_{i+}^2/r) - T'$	$c - 1$	$s_c^2 = S_c/(c-1)$	$F_c = s_c^2/s_R^2$
Row means	$S_r = \sum (T_{+j}^2/c) - T'$	$r - 1$	$s_r^2 = S_r/(r-1)$	$F_r = s_r^2/s_R^2$
Residual	$S_R = S_T - S_c - S_r$	$(c-1)(r-1)$	$s_R^2 = S_R/(c-1)(r-1)$	
Total	$S_T = \sum\sum x_{ij}^2 - T'$	$rc - 1$		

where $T' = T_{++}^2/rc$

column effects, respectively. Similarly, the total sum of squares and d.f. can be calculated as before. In this case, however, it will be found that the sum of the sum of squares and d.f. for the column and row means are less than the total sum of squares and d.f. This deficiency in the sum of squares is attributable to random errors and is assigned to the row labelled "residual" and gives a measure of the variability over and above that attributable to row and column differences. The difference in means for each category can then be tested individually for significance using the ratio of the mean square of each category to that of the mean square of the residuals as shown in Table 3.3.

Numerical example

An example of the two-way classification with single observations is given in Table 3.4 together with the analysis of variance table. The results indicate that at the 95% confidence level there is no evidence to suggest that there is a difference between the means of the four analysts (columns) as $F_{0.05}$ with 3

TABLE 3.4

Example of two-way analysis of variance with single observations using some Sr values (ppm) determined by 4 different analysts on 3 different samples.

Analyst	1	2	3	4	Total
Sample 1	51	47	51	49	198
Sample 2	52	52	58	53	215
Sample 3	50	48	49	48	195
Total	153	147	158	150	608

$S_c = 153^2/3 + 147^2/3 + .. + 150^2/3 - 608^2/12 = 22.0$ $s_c^2 = 22.0/3 = 7.33$

$S_r = 198^2/4 + 215^2/4 + 195^2/4 - 608^2/12 = 58.17$ $s_r^2 = 58.17/2 = 29.09$

$S_T = 51^2 + 47^2 + ... + 49^2 + 48^2 - 608^2/12 = 96.67$

$S_R = 96.67 - 58.17 - 22.0 = 16.50$ $s_R^2 = 16.50/6 = 2.75$

$F_c = 7.33/2.75 = 2.67$ $F_r = 29.09/2.75 = 10.58$

TWO-WAY ANALYSIS OF VARIANCE: SINGLE OBSERVATIONS

Source of variation	Sum of Squares	Degrees of freedom	Variance or mean square	F-ratio
Analysts	22.00	3	7.33	2.67
Samples	58.17	2	29.09	10.58
Residual	16.50	6	2.75	
Total	96.67	11		

Conclusions: Columns (analysts) - $F_c = 2.67$ is not significant as $F_{0.05} = 4.76$ with 3 and 6 d.f.   Rows (samples) - $F_r = 10.58$ is significant as $F_{0.05} = 5.14$ with 2 and 6 d.f.

and 6 d.f. from Table A4 is 4.76, which is greater than the value of the test statistic $F_c = 2.67$. There is, however, evidence to suggest that there is a significant difference in the mean of the samples (rows) as $F_{0.05}$ with 2 and 6 d.f. is 5.14, which is less than $F_r = 10.58$.

3.4 TWO-WAY CLASSIFICATION WITH MULTIPLE OBSERVATIONS

This design is similar to the previous one except that multiple observations are made for each pair of categories. Although methods are available for dealing with different numbers of observations in each cell, only the simplest case where each cell has an equal number of observations will be considered here. The advantage of this design over the previous one is that it provides a better estimate of the experimental errors and also enables the data to be assessed more fully. This design has been used extensively in interpreting the results of the eight new USGS rock standards (Flanagan, 1976a).

The generalized computational scheme for this design is given in Table 3.5. The sum of squares of the total, column and row means are obtained as in the previous design. However, in this case as we have multiple observations in each cell we can obtain an additional estimate of the variance S_S, using the totals of each cell. Note that in the previous design this estimate is the same as the total sum of squares. The difference between the total sum of squares S_T and S_S is called the sum of squares of interaction S_I and is equivalent to the residual sum of squares in the previous design. The interaction is an additional component which is assumed to be present in the mean of each cell and can be different for each cell. Each of the sources of variation can then be tested for significance by comparing its mean square with that of the within groups mean square s_w^2.

If the interaction effect is not significant, it is permissible to add the sum of squares of the interaction and within groups together, in order to give a better estimate of the population variance, which can then be used to form new F-ratios to test the column and row means as shown at the bottom of Table 3.5.

Numerical example

The example given in Table 3.6 is taken from Flanagan (1976a) and consists of the results of duplicate Be determinations made by three different laboratories on three splits or thirds of the USGS standard STM-1. The analysis of variance indicates that the interaction effect is not significant at the 95% confidence level, as $F_{0.05} = 3.63$ with 4 and 9 d.f., which is greater than the test statistic $F_I = 1.74$. Similarly, at the 95% confidence level there is no evidence to suggest that there are differences between the means of the different thirds, but there is evidence to suggest that there are significant differences between the means of the laboratories, as the value of $F_{0.05}$ with 2 and 9 d.f. is 4.26, which is less than the test statistic $F_c = 6.89$.

TABLE 3.5

Generalized computational scheme for a two-way analysis of variance with n repeated observations in each cell. T_{ij+} is the sum of the n observations for the ith category 1 and the jth category 2. x_{ijk} is the kth observation of the ith category 1 and jth category 2.

| Category 2 | Category 1 | | | | | |
	1	2	3	.	c	Row Totals
1	T_{11+}	T_{21+}	T_{31+}	...	T_{c1+}	T_{+1+}
2	T_{12+}	T_{22+}	T_{32+}	...	T_{c2+}	T_{+2+}
3	T_{13+}	T_{23+}	T_{33+}	...	T_{c3+}	T_{+3+}
.	...	...	...	...	...	...
r	T_{1r+}	T_{2r+}	T_{3r+}	...	T_{cr+}	T_{+r+}
Column Totals	T_{1++}	T_{2++}	T_{3++}	...	T_{c++}	T_{+++}

TWO-WAY ANALYSIS OF VARIANCE: MULTIPLE OBSERVATIONS

Source of variation	Sum of Squares	Degrees of freedom	Variance or mean square	F-ratio[*]
Col means	$S_c = \sum (T_{+j+}^2/nc) - T'$	$c - 1$	$s_c^2 = S_c/(c-1)$	$F_c = s_c^2/s_w^2$
Row means	$S_r = \sum (T_{i++}^2/nr) - T'$	$r - 1$	$s_r^2 = S_r/(r-1)$	$F_r = s_r^2/s_w^2$
Interaction	$S_I = S_S - S_c - S_r$	$(c-1)(r-1)$	$s_I^2 = S_I/(c-1)(r-1)$	$F_I = s_I^2/s_w^2$
Sub-total	$S_S = \sum\sum (T_{ij+}^2/n) - T'$	$rc - 1$		
Within	$S_w = S_T - S_S$	$rc(n-1)$	$s_w^2 = S_w/rc(n-1)$	
Total	$S_T = \sum\sum\sum x_{ijk}^2 - T'$	$nrc - 1$		

where $T' = T_{+++}^2/nrc$
(*) S_I and S_w can be pooled to give $s_R^2 = (S_I+S_w)/(nrc-r-c+1)$, if F_I is not significant. The F-ratios then become $F_c' = s_c^2/s_R^2$ and $F_r' = s_r^2/s_R^2$ with $nrc-r-c+1$ d.f. for the denominator.

In this example, pooling the interaction and within groups sum of squares does not effect the result as neither are borderline cases. Remember that when pooling these two variances, not only do the F-ratios change, but the d.f. of the denominators also change.

3.5 HOMOGENEITY OF VARIANCES

An underlying assumption of the analysis of variance designs is that the populations all have equal variances. Slight departures from this assumption cause little trouble if the samples are all of equal size, but problems can arise

TABLE 3.6

Example of two-way analysis of variance with duplicated observations, using Be determinations reported on the USGS standard STM-1 (Flanagan, 1976a) in which samples from three different splits of the standard (thirds) were analysed in duplicate by three different laboratories.

Thirds	Laboratories								Total
	A		Sub-total	B		Sub-total	C		Sub-total	
1	5	6	11	10	11	21	7	14	21	53
2	6	6	12	17	12	29	11	10	21	62
3	5	7	12	7	6	13	14	8	22	47
Total			35			63			64	162

$T' = 162^2/(2 \times 3 \times 3) = 1458$ $S_T = 5^2 + 10^2 + 7^2 + .. + 6^2 + 8^2 - 1458 = 214$

$S_c = 35^2/6 + 63^2/6 + 64^2/6 - 1458 = 90.33$ $s_c^2 = 90.33/2 = 45.17$

$S_r = 53^2/6 + 62^2/6 + 47^2/6 - 1458 = 19.0$ $s_r^2 = 19.0/2 = 9.5$

$S_S = 11^2/2 + 21^2/2 + 21^2/2 + ... + 22^2/2 - 1458 = 155$

$S_I = 155.0 - 19.0 - 90.33 = 45.67$ $s_I^2 = 45.67/4 = 11.42$

$S_W = 214.0 - 155.0 = 59.0$ $s_W^2 = 59.0/9 = 6.56$

$F_c = 45.17/6.56 = 6.89$ $F_r = 9.5/6.56 = 1.45$

$F_I = 11.42/6.56 = 1.74$ $s_R^2 = (45.67 + 59.0)/(4 + 9) = 8.05$

$F_c' = 45.17/8.05 = 5.61$ $F_r' = 9.5/8.05 = 1.18$

TWO-WAY ANALYSIS OF VARIANCE: MULTIPLE OBSERVATIONS

Source of variation	Sum of Squares	Degrees of freedom	Variance or mean square	F-ratio
Laboratories	90.33	2	45.17	6.89
Thirds	19.00	2	9.50	1.45
Interaction	45.67	4	11.42	1.74
Sub-total	155.00	8		
Within	59.00	9	6.56	
Total	215.00	17		

Conclusions: Interaction - $F_I = 1.74$ is not significant as $F_{0.05} = 3.63$ with 4 and 9 d.f. Columns (Labs) - $F_c = 6.84$ is significant as $F_{0.05} = 4.26$ with 2 and 9 d.f. Rows (Thirds) - $F_r = 1.45$ is not significant as $F_{0.05} = 4.26$ with 2 and 9 d.f.

if the samples are of unequal size, or if one of the variances is much larger than the others. The usual method for testing the hypothesis that all the variances are equal is called *Bartlett's test*, which produces a test statistic which closely follows the χ^2 distribution. Assuming that we have k samples of size $n_1, n_2, \ldots n_k$, with variances $s_1{}^2, s_2{}^2, \ldots s_k{}^2$, we calculate the pooled variance $s_p{}^2$ as:-

$$s_p{}^2 = \sum (n_i - 1) \, s_i{}^2 \, / \, (N - k) \tag{3.1}$$

where $N = \sum n_i$. We then calculate the two statistics M and C as:-

$$M = (N - k) \log_e s_p{}^2 - \sum (n_i - 1) \log_e s_i{}^2 \tag{3.2}$$

and

$$C = 1 + \frac{1}{3(k-1)} \left[\sum \frac{1}{n_i - 1} - \frac{1}{N - k} \right] \tag{3.3}$$

to form:-

$$\chi^2 = M \, / \, C \quad \text{with } k - 1 \text{ d.f.} \tag{3.4}$$

Unfortunately, this test is very sensitive to data that are not normal, when it tends to reject the null hypothesis too frequently. Special procedures should also be adopted if the number of observations in most of the groups is less than 5 (Snedecor & Cochran, 1971, p.298).

Another test, called *Cochran's test* is also available to test if one variance is significantly larger than the rest. It is based on the statistic G where:-

$$G = \text{largest } s_i{}^2 \, / \, \sum s_i{}^2 \tag{3.5}$$

The critical values of G for particular levels of significance can be found in special tables (Dixon & Massey, 1969, Table A-17; Walpole & Myers, 1972, Table XI) and are entered with the number of d.f. of the largest variance and k, the number of groups.

If as a result of these tests it is suspected that the variances are not equal, the correct procedure is to try to transform the data to stabilize the variances, as discussed in section 1.3.5.

Numerical example

A worked example of these methods is given in Table 3.7, using the data of Table 3.2. Bartlett's test gives a value of $\chi^2 = 7.85$, so that the null hypothesis of equal variances is rejected as $\chi^2{}_{0.05} = 5.99$ with 2 d.f. Cochran's test, on the other hand indicates that the largest variance (607.2) is not significantly larger than the other two as $G_{0.05} = 0.7977$ for 3 groups and 4 d.f., which is greater than the calculated value of 0.5803.

TABLE 3.7

Example of Bartlett's test and Cochran's test for the homogeneity of variance using the Sr data determined by 3 different analytical methods from Table 3.2.

	Optical spectrographic	Atomic absorption	X-ray fluorescence
Variance	607.2	367.6	71.6061
Number obs.	5	6	12

Bartlett's test

$s_p^2 = (4 \times 607.2 + 5 \times 367.6 + 11 \times 71.6061) / (23 - 3) = 252.72$ (eqn 3.1)

$M = 20\log_e 252.72 - 4\log_e 607.2 - 5\log_e 367.6 - 11\log_e 71.6061 = 8.49$ (eqn. 3.2)

$C = 1 + (1/4 + 1/5 + 1/11 - 1/20)/6 = 1.0818$ (eqn. 3.3)

$\chi^2 = 8.49/1.0818 = 7.85$ with 2 d.f.   (eqn. 3.4)

Cochran's test

$G = 607.2/(607.2 + 367.6 + 71.6061) = 0.5803$ (eqn. 3.5)

CHAPTER 4

LINEAR RELATIONSHIPS BETWEEN TWO VARIABLES

4.1 INTRODUCTION

So far we have been dealing with problems involving one variable only, but most data involve simultaneous measurements on more than one variable. Such data are known as *multivariate* and much of the remainder of the book will be dealing with this type of data.

In petrology the variation diagram is still one of the commonest ways of interpreting data, and this chapter is concerned with three ways of evaluating linear trends within such bivariate plots. The methods are the use of the correlation coefficient, linear regression analysis, and either functional or structural relationships. The choice of method basically depends upon the object of investigating the relationship between the two variables. If all that is required is to establish that a linear relationship exists between the two variables, then the *correlation coefficient* should be used and tested for significance. However, if a knowledge of the parameters that control the variation is important, then either *functional* or *structural relationship* should be used. This is often the case with variation diagrams, where the petrologist has some preconceived ideas about the direction of variation e.g. as with olivine control lines. *Linear regression analysis*, on the other hand, should be confined to those problems where it is desired to predict the value of one variable from knowledge of others. Mark & Church (1977) and Jones (1979) have presented excellent discussions on the subtle differences between the regression approach and that of functional and structural relationships and point out that regression analysis is often misused in the Earth Sciences. However, as the absolute value of the correlation coefficient approaches unity, the difference between the two methods becomes trivial, so that under these circumstances the correct choice of method is largely academic.

4.2 THE CORRELATION COEFFICIENT

The correlation coefficient is the simplest measure of the linear relationship between two variables. Just as a single variable has a frequency distribution, so two or more variables can have joint frequency distributions, which for reasons previously stated are usually assumed to be normal (section 1.3.1). A representation of a simple bivariate normal distribution showing contours of equal probability is given in Fig. 4.1. Although a univariate frequency distribution can be defined entirely in terms of a mean and a variance, a bivariate frequency distribution has to be defined not only by the means and variances of the two variables, but also with a new quantity called the *covariance*, which is defined as follows:-

$$\text{cov}_{12} = \sum(x_{i1} - \bar{x}_1)(x_{i2} - \bar{x}_2) / (n - 1) = SP_{12} / (n - 1) \qquad (4.1)$$

where cov_{12} is the covariance between variable 1 and 2, x_{i1} and x_{i2} are the values of variables 1 and 2 for the ith object, $\bar{x}_1$ and $\bar{x}_2$ are the sample means of the two variables, and n is the number of pairs of observations. Note that if the two variables are the same, then the covariance simply becomes the variance of the variable as written in equation 2.2.

The numerator of equation 4.1 is a *sum of products* of deviations from two means and is again a commonly used expression in statistics and will be abbreviated SP (Sum of Products) with two subscripts to indicate the variables involved. By analogy with the manipulation of the sum of squares (equation 2.3) it can be shown that:-

$$SP_{12} = \sum(x_{i1} - \bar{x}_1)(x_{i2} - \bar{x}_2)$$
$$= \sum x_{i1}x_{i2} - \left[(\sum x_{i1})(\sum x_{i2}) / n\right] \qquad (4.2)$$

so that the covariance can also be expressed in a more convenient

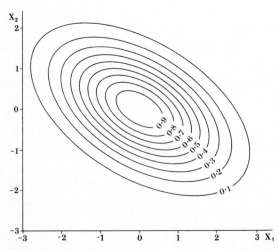

Fig. 4.1. An example of a bivariate normal distribution between two variables x_1 and x_2, when the correlation coefficient r is -0.5 and the standard deviations of x_1 and x_2 are 1.3 and 1.0, respectively. The family of ellipses, which are contours of equal probability, are numbered with the proportion of the distribution that falls outside each ellipse. Any section through the distribution, parallel to the x_1 or x_2 axis, will produce a normal frequency distribution as shown in Fig. 1.2.

computational form as :-

$$cov_{12} = \left[\sum x_{i1}x_{i2} - (\sum x_{i1})(\sum x_{i2}) / n\right] / (n - 1) \tag{4.3}$$

Inspection of equation 4.1 shows that if x_{i1} and x_{i2} both tend to increase in value together, then $x_{i1} - \bar{x}_1$ and $x_{i2} - \bar{x}_2$ will both tend to be of the same sign more often than of opposite sign, so that the covariance will be positive. This relationship is called a *positive correlation* as opposed to a *negative correlation* when one variable increases as the other decreases.

The covariance, however, is a difficult measure to interpret as its value not only depends upon the units of measurement of the two variables, but it may also be the product of two completely different types of unit e.g. a distance times a weight.

It is for this reason that the *correlation coefficient*, r, is used. This is defined as:-

$$r_{12} = cov_{12} / \sqrt{(var_1 \times var_2)} \tag{4.4}$$

where var_1 and var_2 are the variances of the two variables. A new nomenclature for variance is introduced here, as it is easier to use in the multivariate examples later in the book than the s_i^2 nomenclature as used in Chapter 2. The correlation coefficient can, of course, also be written as:-

$$r_{12} = cov_{12} / (s_1 \times s_2) \tag{4.5}$$

where s_1 and s_2 are the standard deviations of the two variables, or:-

$$r_{12} = SP_{12} / \sqrt{(SS_1 \times SS_2)} \tag{4.6}$$

To see what correlation coefficients mean in terms of the scatter of data points, some hypothetical sets of data have been plotted in Fig. 4.2. At a more generalized level, the way in which the shape of the bivariate normal distribution is controlled by the standard deviations of the two variables and the correlation coefficient is illustrated in Fig. 4.3. Several important points should be noted:-

1. The correlation coefficient is a unitless quantity with a range between -1 and 1.
2. The closer r is to 1 or -1, the closer the data lie to a straight line.
3. This straight line *does not* have to pass through the origin e.g. Fig. 4.2a.
4. A correlation coefficient of zero only implies that there is no linear relationship between the variables. It does not imply that there is no other relationship between them e.g. a curvilinear relationship as in Fig. 4.2f.
5. If the standard deviations of the two variables are held constant, the family of ellipses of equal probability inscribe a rectangle whose sides are proportional to the two standard deviations (Fig. 4.3a).
6. The angles the axes of an ellipse of equal probability make with the variable axes, depend not only upon the correlation coefficient, but also upon the standard deviations of the variables, as can be seen in Figs. 4.3a and 4.3b.

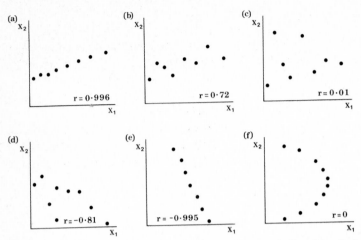

Fig. 4.2. Some diagrams to illustrate the relationship between the value of correlation coefficients and the scatter of points. Note that a zero correlation coefficient only means that there is no linear relationship between the variables.

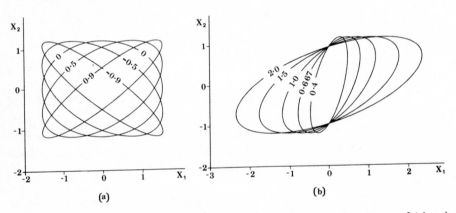

Fig. 4.3. Diagrams to illustrate the factors controlling the shape of bivariate normal distributions. In both diagrams, each ellipse defines the area within which half of the distribution lies. In (a) the standard deviations of x_1 and x_2 remain constant at 1.3 and 1.0, respectively, as the correlation coefficient r changes as shown on the ellipses; in (b) the standard deviation of x_1 changes as shown on the ellipses, as the correlation coefficient r and the standard deviation of x_2 remain constant at 0.6 and 1.0, respectively.

Numerical example

A worked example of the calculation of a correlation coefficient between the depth in kilometres to the underlying Benioff zone (y) and the value of the anhydrous cationic K% when Si% = 51 (x), for volcanic rocks from 17 volcanoes in Java (Hutchison, 1975), is given in Table 4.1.

TABLE 4.1

Worked example of the calculation of a correlation coefficient between the depth to the underlying Benioff zone (y), and the anhydrous cationic K% of lavas at Si% = 51 (x), using data from 17 volcanoes in Java (Hutchison, 1975, Table 1, p.162) which will be assumed to be a random sample. The location of the volcanoes can be seen in Fig. 6.1.

Name of volcano	Depth km (y)	K% at Si% = 51 (x)
1. Krakatau	140	1.62
2. Tangkuban	180	0.98
3. Papandajan	140	0.24
4. Guntur	155	1.08
5. Galunggung	150	0.71
6. Tjerimai	190	1.37
7. Slamet	185	1.74
8. Dieng	200	2.39
9. Sumbing	185	0.86
10. Ungaran	215	3.27
11. Merbabu	185	1.95
12. Merapi	175	2.29
13. Kelud	170	1.23
14. Semeru	165	1.38
15. Bromo	180	2.15
16. Raung	190	2.18
17. Batur	195	1.65

$\sum y_i = 3000.0$ $\sum x_i = 27.09$

$\sum y_i^2 = 536400.0$ $\sum x_i^2 = 51.9233$

$n = 17$ $\sum x_i y_i = 4954.3$

$\bar{y} = 176.4706$ $\bar{x} = 1.5935$

$SP_{xy} = 4954.3 - 3000.0 \times 27.09 / 17 = 173.7118$ (eqn. 4.2)

$SS_x = 51.9233 - 27.09^2 / 17 = 8.7546$ (eqn. 2.3)

$SS_y = 536400.0 - 3000.0^2 / 17 = 6988.2353$ (eqn. 2.3)

$r_{xy} = 173.7118 / \sqrt{(8.7546 \times 6988.2353)} = 0.70$ (eqn. 4.6)

4.2.1 *Testing the correlation coefficient for significance*

The linear dependence of two variables can be tested by setting up the null hypothesis that the population correlation coefficient, ρ (Greek letter rho), is zero. If this hypothesis is rejected it means that there is evidence that the two variables are linearly dependent upon each other. The test, which uses the correlation coefficient, r, calculated from a *random sample* of n observations, is based upon the relationship that the variable t, defined as:-

$$t = r \sqrt{\left[(n - 2) / (1 - r^2) \right]}$$

(4.7)

is distributed as Student's t with (n - 2) d.f. and can be tested against values of $t_{\alpha/2}$ obtained from Table A3. Rewriting this as:-

$$r = t_{\alpha/2} / \sqrt{(t_{\alpha/2}^2 + n - 2)}$$

(4.8)

enables Table A6 in the Appendix to be computed, which can be used to give values of r which must be exceeded to reject the null hypothesis at different confidence levels. In making this test, the sign of r may be ignored as the distribtuion of r is symmetrical when $\rho = 0$.

Numerical example

Assuming that the depth to the Benioff zone versus K% data of Table 4.1 are a random sample, we will see if, at a confidence level of 95%, the sample correlation coefficient of 0.7023 could have come from a population with $\rho = 0$. Using equation 4.7 we get:-

$$t_{0.025} = 0.7023 \sqrt{\left[15 / (1 - 0.7023^2) \right]} = 3.82 \quad \text{with 15 d.f.}$$

As the critical value of $t_{0.025}$ with 15 d.f. is 2.131 (Table A3) the null hypothesis is rejected and we conclude that such a sample could not have come from a population with a zero correlation coefficient. Alternatively, Table A6 gives r = 0.482 for 17 observations (d.f. = 15) which, being less than the sample value of 0.7023, means that the null hypothesis is rejected, at the 95% confidence level.

4.2.2 Confidence limits for the population correlation coefficient

The previous test can only be used to test the null hypothesis that $\rho = 0$. If it is required to test the hypothesis that ρ equals a particular value, confidence limits can be used (e.g. Snedecor & Cochran, 1971, p.185). These may be calculated by using the fact that the variable z_r, which is given by:-

$$z_r = 0.5 \log_e \frac{1 + r}{1 - r} \tag{4.9}$$

is, except for extremely small samples, approximately normally distributed with mean z_ρ and standard deviation $1 / \sqrt{(n - 3)}$, where n is the number of pairs of observations used in calculating r. By analogy with equation 2.10, the $100(1 - \alpha)$% confidence limits are then given by:-

$$z_r - z_{\alpha/2} / \sqrt{(n - 3)} \leqslant z_\rho \leqslant z_r + z_{\alpha/2} / \sqrt{(n - 3)} \tag{4.10}$$

where $z_{\alpha/2}$ is the standard normal deviate obtained from Table A2. The value of z_r can either be calculated from equation 4.9 or obtained from Table A7 in the Appendix. Once the confidence limits have been calculated they can be converted back to correlation coefficients either by rewriting equation 4.9 which gives r as:-

$$r = \frac{e^{2z} - 1}{e^{2z} + 1} \tag{4.11}$$

or by using Table A7.

Numerical example

Assuming that the data of Table 4.1 are a random sample, we will determine the confidence limits for the population correlation coefficient at the 95% confidence level. Using equation 4.9 we obtain:-

$$z_r = 0.5 \log_e(1.7023 / 0.2977) = 0.872$$

so that using equation 4.10 we have:-

$$0.872 - 1.96 / \sqrt{14} \leqslant z_\rho \leqslant 0.872 + 1.96 / \sqrt{14}$$

which gives:-

$$0.348 \leqslant z_\rho \leqslant 1.396$$

Finally, using equation 4.11 or Table A7 to convert these limits back to correlation coefficients we get the 95% confidence limits as:-

$$0.335 \leqslant \rho \leqslant 0.884$$

In other words, we can be 95% confident that the population correlation coefficient, ρ, has a value between 0.335 and 0.884.

4.2.3 *The problem of closure*

In a chemical analysis of a rock, SiO_2 cannot increase without some of the other oxides decreasing, due to the fact that the analysis must sum to 100%. This property is known as *closure*. To be more specific, it is one particular type of closure known as the constant sum type and is displayed by all major element data. In general, closure is the property whereby given $n - 1$ variables from a set of data, the nth is determined. Another type of closure is that displayed by cation values in mineral recalculations.

As first emphasized by Chayes (1960, 1962), closure invalidates the use of the usual null hypothesis that the population correlation coefficient is zero, as a test to see if the two variables are independent (section 4.2.1). He showed that for any set of constant sum data, the sum of the covariances of any oxide with all the others is equal to the negative value of the variance of the particular oxide. This can be written as:-

$$\sum_{j=1}^{n} cov_{ij} = -var_i \quad \text{with } j \neq i \tag{4.12}$$

This means that the correlation coefficients will be far more negative than would be expected for ordinary data, so that to be considered significant, a negative correlation coefficient would have to be much more negative than suggested by the test of section 4.2.1. Similarly, a positive correlation coefficient may be more significant than the test suggests. Unless this point is realized, it is easy to be trapped into ascribing petrological significance to trends in variation diagrams that may be due entirely to closure (see also Butler, 1979a).

In an attempt to solve this problem Chayes & Kruskal (1966) use the concept of a theoretical open array of data with zero covariances between all variables, such that when closed it generates the variances and covariances observed between all the variables. The expected values of the correlation coefficients in the closed data can then be calculated, so that significance tests can be applied as shown in the previous section.

However, even this approach is not without its problems. For petrological data the concept of an open array has little physical meaning, as closure is a function of the objects we deal with and not a function of the measurements. For example, just as Mg and Fe must vary inversely in an olivine as they are competing for space within the structure, so olivine must vary inversely with pyroxene in peridotites, where the two minerals are competing for space within the rock. As Miesch (1969, p.174) states "The fundemental cause of the constant sum problem in geochemistry is the fact that the amount of chemical constituents in a rock sample is a meaningless variable. It is necessary to measure the amount in relationship to the amounts of one or more other constituents ... ".

Another problem is that there are also an infinite number of open arrays, all with different properties, that will close to give a particular observed array. This can easily be seen geometrically with reference to Fig. 5.1, where closed data lying entirely in the Or–Ab–An plane can be opened by projecting each data point *any* distance towards or away from the origin. As the open array has little physical meaning, there seems to be little justification for assuming that one particular open array (as chosen by Chayes & Kruskal, 1966) is more relevant than any other.

Furthermore, both Chayes (1971) and Butler (1975) have shown that for some sets of data the Chayes-Kruskal procedure produces an open array with negative variances which is, of course, a contradiction in terms. For such data Chayes has suggested that the null hypothesis should be automatically rejected. However, as these negative variances result directly from the assumptions made about the open data, it could also mean that some of the assumptions are incorrect or even that the whole concept of the open array for petrological data is suspect. Kork (1977) also showed that if the null hypothesis of Chayes & Kruskal (1966) is rejected, the method is incapable of deciding which correlations are significant which, of course, is one of the major points of interest.

An extremely promising new approach has recently been used by Aitchison (1981) who points out that one of the major problems with the conventional approach is that the data lie in a simplex (a restricted region) in p-dimensional space, which makes the mathematics of significance testing extremely difficult. However, by applying a log-ratio transformation to the original data, one obtains a new set of data which can be considered mathematically to lie in a much less restricted region in p-dimensional space to which the multivariate normal distribution can be applied. A computer program is also available for this procedure from the author.

Although many other papers have appeared on the problems of closure and correlation (e.g. Chayes, 1964, 1967, 1970b; Darroch, 1969; Darroch & Ratcliff, 1970, 1978; Miesch et al., 1966; Saha et al., 1974; Skala, 1977, 1979; Snow, 1975; Vistelius & Sarmanov, 1961; Zodrow, 1976) some authors still persist in using the method of section 4.2.2 to test the significance of correlation coefficients calculated from major element data. This may be due to ignorance or as Till (1974, p.91) states "The mathematics involved is complex and this leads most geologists to ignore the work". If either of these are true, it is a sad reflection on the geological profession! It is, therefore, important to realize that there is a problem with significance tests for correlation coefficients calculated from closed data. However, if the correlation coefficient is used only as a sample descriptor, no problems will arise.

Finally, it should be noted that the problem cannot be avoided by not recalculating the analysis to 100%, as the difference between the actual total and 100% can be regarded as an undetermined variable e.g."others". Similarly, recalulation to cations, molecular proportions etc. does not help, for as long as the initial data are closed, no amount of recalculation will open them. The effect of recalculations on the closure problem has been further discussed in great detail by Butler (1979b, 1980).

4.3 LINEAR REGRESSION ANALYSIS

The classic method of linear regression analysis, which is described in great detail in many excellent practical texts e.g., Draper & Smith (1966) and Williams (1959), should be confined to problems where it is desired to predict the value of one variable, called the *dependent variable*, from knowledge of the value of another variable, called the *independent variable*. The method usually assumes that the independent variable either has assigned values or can be measured with neglible error, while the dependent variable is subject only to random errors. Although this would appear to limit the choice of independent variable, it has been shown (Lindley, 1947) that linear regression is valid even if the independent variable is subject to error, as long as the values of the independent variable used for the prediction, come from the same population as that used for estimating the regression. For example, suppose that a series of lavas are thought to have a linear relationship between SiO_2 and K_2O. Regression analysis will theoretically enable K_2O to be correctly predicted from *observed* values of SiO_2 (measured with errors and from the same population), but will give biased predictions of K_2O at specified values of SiO_2 (known without error).

In linear regression the underlying mathematical model for the population is that the dependent variable y is related to the independent variable x by an equation of the form:-

$$y = \alpha + \beta x + \varepsilon \qquad (4.13)$$

where α and β (Greek letter beta) are the population regression constants and ε (Greek letter epsilon) is a random variable, independent of the value of x, with zero mean and variance $\sigma_{y.x}^2$ (see equation 4.26). This model is called a *regression of Y on X*. One of the objects of regression analysis is to provide estimates of α and β, which will be denoted a and b, so that we may write:-

$$\hat{y} = a + bx \qquad (4.14)$$

where $\hat{y}$ (read y hat) is the estimated value of y for a given value of x. Now for each of the n pairs of randomly selected observations we can write:-

$$y_i = \alpha + \beta x_i + \varepsilon_i \qquad (4.15)$$

where ε_i is the random deviation of the ith pair of observations x_i, y_i, from the regression line. This is shown diagramatically in Fig. 4.4. To be able to estimate a and b we must first decide which criterion to use in fitting the line. For example, we could fit the line to minimize the sum of the absolute values of ε_i, or the sum of the signed values of ε_i. However, both of these methods give results that are not entirely satisfactory (e.g. see Wonnacott & Wonnacott, 1977). The method usually chosen is the *method of least-squares*, which minimizes the sum of squares of the deviations from the regression line. The quantity to be minimized is therefore:-

$$S = \sum \varepsilon_i^2 = \sum (y_i - \alpha - \beta x_i)^2 \qquad (4.16)$$

This is done by differentiating S with respect to α and β and equating the

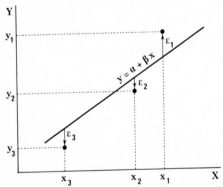

Fig. 4.4. Diagram to illustrate the way in which a regression line of the form y = a + bx, is normally fitted to a set of data points, by minimizing the sum of squares of the deviations of y from the regression line, i.e. the sum of $\varepsilon_1^2 + \varepsilon_2^2 + \varepsilon_3^2 + \ldots$ etc.

results to zero, which gives:-

$$\frac{\partial S}{\partial \alpha} = -2 \sum (y_i - \alpha - \beta x_i)$$

$$\frac{\partial S}{\partial \beta} = -2 \sum x_i(y_i - \alpha - \beta x_i) \tag{4.17}$$

Substituting a and b, the estimated values of α and β, then gives:-

$$\sum (y_i - a - bx_i) = 0$$

$$\sum x_i(y_i - a - bx_i) = 0 \tag{4.18}$$

By expanding and rearranging these equations, we then obtain what are known as the *normal equations*:-

$$an + b\sum x_i = \sum y_i$$

$$a\sum x_i + b\sum x_i^2 = \sum x_i y_i \tag{4.19}$$

where n is the number of pairs of observations. Solving these for b, the slope of the line we obtain:-

$$b = \left[\sum x_i y_i - (\sum x_i)(\sum y_i) / n\right] / \left[\sum x_i^2 - (\sum x_i)^2 / n\right]$$

$$= SP_{xy} / SS_x \tag{4.20}$$

using the abbreviations of equations 4.2 and 2.3. Equation 4.14 then gives a, the intercept as:-

$$a = \bar{y} - b\bar{x} \tag{4.21}$$

where $\bar{y}$ and $\bar{x}$ are the mean values of the variables.

Note that in general, the line obtained from a regression of Y on X will be different from the line obtained from a regression of X on Y. If b and b' are the slopes of the two regression lines $y = a + bx$ and $x = a' + b'y$, it follows that they will only define the same line if $bb' = 1$. But by using equations 4.20 and 4.6, it follows that:-

$$bb' = SP_{xy}^2 / SS_x SS_y = r_{xy}^2 \tag{4.22}$$

so that the two lines are only equal when the correlation coefficient is 1 or -1. When $r = 0$ the two lines are at right angles to each other.

Numerical example

Again using the data of Table 4.1, we can calculate the coefficient and intercept for a regression of Y on X from equations 4.20 and 4.21 as:-

b = 173.7118 / 8.7546 = 19.8423   and
a = 176.4706 - 19.8423 x 1.5935 = 144.8519

It is then possible to estimate the depth to the underlying Benioff zone (y) from

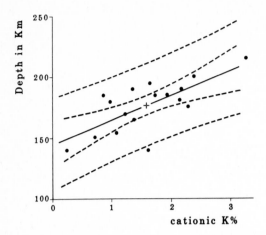

Fig. 4.5. Plot of the depth in kilometres to the underlying Benioff zone (y), and the anhydrous cationic K% of lavas at Si% = 51 (x), using data from 17 volcanoes in Java (Hutchison, 1975). The line for the regression of Y on X is shown, together with the 95% confidence limits for the regression line (the inner pair of curves) and individual values of y (the outer pair of curves).

the equation y = 144.85 + 19.84 x K%. When K% is 3.0, the estimated depth to the underlying Benioff zone is, therefore, given by:-

Depth = 144.85 + 19.84 x 3.0 = 204 km.

A plot of all the data points from Table 4.1 and this regression line is shown in Fig. 4.5 above.

4.3.1 *Testing the significance of a regression*

For linear regression, tests for the "significance of a regression" are equivalent to testing the null hypothesis that the slope of the regression line is zero. This is often performed by an analysis of variance, in which the total sum of squares about the mean is divided into two components, one due to the regression and the other due to the errors or departures from the regression line. Now, the total sum of squares about the mean, S_T, which has n – 1 d.f., is given by:-

$$S_T = \sum y_i^2 - (\sum y_i)^2 / n = SS_y \qquad (4.23)$$

and the sum of squares due to the regression, S_R, which has 1 d.f., is given by:-

$$S_R = b\left[\sum x_i y_i - (\sum x_i)(\sum y_i) / n\right] = b\, SP_{xy} \qquad (4.24)$$

where b is the estimated regression coefficient. The sum of squares of the errors or deviations from the regression line, S_E, which has n – 2 d.f., is then given by the difference:-

$$S_E = SS_y - b\, SP_{xy} \qquad (4.25)$$

Of particular importance is the error mean square, $s_{y.x}^2$, where:-

$$s_{y.x}^2 = (SS_y - b\,SP_{xy}) / (n - 2) \qquad\qquad (4.26)$$

which is an unbiased estimate of $\sigma_{y.x}^2$, the population variance of ε in equation 4.13, and is used in setting confidence limits. The generalized analysis of variance is given in Table 4.2 and if the F-ratio exceeds the critical value of F with 1 and $n - 2$ d.f. as given in Table A4 in the Appendix, the null hypothesis is rejected i.e. the slope of the linear regression line is significantly different from zero.

As a check on the calculations, note that the square of the correlation coefficient is the proportion of the total sum of squares that is accounted for by the regression, i.e $S_R/S_T = r^2$. This is simply a restatement of equation 4.22.

If both variables used in the linear regression are from closed data, care must be exercised in interpreting the meaning of a significant regression, just as was done for correlation coefficients. All that can be stated is that the regression is *numerically* significant and is, therefore, suitable for predictive purposes. This does not mean, however, that the regression is *petrologically* meaningful, as the significance may have been produced largely by the closure of the data, and not by any petrological process.

TABLE 4.2

Generalized computational scheme for the analysis of variance for testing the significance of a linear regression, together with a numerical example using the depth to the Benioff zone versus K% data from Table 4.1, which are assumed to be a random sample.

ANALYSIS OF VARIANCE: LINEAR REGRESSION

Source of variation	Sum of squares	Degrees of freedom	Variance or mean square	F-ratio
Regression	$S_R = b\,SP_{xy}$	1	$s_r^2 = S_R$	$s_r^2 / s_{y.x}^2$
Error	$S_E = SS_y - b\,SP_{xy}$	$n - 2$	$s_{y.x}^2 = S_E / (n - 2)$	
Total	$S_T = SS_y$	$n - 1$		

From Table 4.1 we have $SP_{xy} = 173.7118$ and $SS_y = 6988.2353$, while from the previous example we found $b = 19.8423$, so that the sum of squares due to the regression, $S_R = b\,SP_{xy} = 19.8423 \times 173.7118 = 3446.8416$.

NUMERICAL EXAMPLE

Source of variation	Sum of squares	Degrees of freedom	Variance or mean square	F-ratio
Regression	3446.8416	1	3446.8416	14.6
Errors	3541.3937	15	236.0929	
Total	6988.2353	16		

Numerical example

A worked example of the analysis of variance for the depth to the Benioff zone versus K% data of Table 4.1, is also given in Table 4.2. Assuming that the data are a random sample, the linear regression is highly significant, as the F-ratio of 14.6 with 1 and 15 d.f. greatly exceeds the critical value of 8.68.

4.3.2 *Confidence limits for the regression coefficient and intercept*

The variance of b, the slope of the regression line, is given by the expression:-

$$s_b^2 = s_{y.x}^2 / SS_x \tag{4.27}$$

so that the $100(1 - \alpha)\%$ confidence limits on the slope of the population regression line are given by:-

$$b \pm t_{\alpha/2}\, s_b \tag{4.28}$$

where $t_{\alpha/2}$ is the value of Student's t with n - 2 d.f. obtained from Table A3 in the Appendix. Similarly, the variance of a, the intercept of the regression line is given by the expression:-

$$s_a^2 = s_{y.x}^2 \sum x_i^2 / nSS_x \tag{4.29}$$

so that the $100(1 - \alpha)\%$ confidence limits on the intercept of the population regression line are given by:-

$$a \pm t_{\alpha/2}\, s_a \tag{4.30}$$

where $t_{\alpha/2}$ is the same as before.

Numerical example

Continuing to use the depth to the Benioff zone versus K% data of Table 4.1, we have from Table 4.2 that $s_{y.x}^2 = 236.0929$, so that equation 4.27 then gives:-

$$s_b^2 = 236.0929 / 8.7546 \quad \text{or} \quad s_b = 5.1931$$

The 95% confidence limits on the slope of the regression line are then given by equation 4.28 as:-

$$19.8423 \pm 2.131 \times 5.1931 \quad \text{or} \quad 20 \pm 11$$

Similarly, to obtain confidence limits for the intercept equation 4.29 gives:-

$$s_a^2 = (236.0929 \times 27.09) / (17 \times 8.7546) \quad \text{or} \quad s_a = 6.5555$$

which gives the 95% confidence limits for a as:-

$$144.8519 \pm 2.131 \times 6.5555 \quad \text{or} \quad 145 \pm 14$$

4.3.3 *Confidence limits for the location of the regression line*

It can be shown (e.g. Draper & Smith, 1966, p.22) that the variance of $\hat{y}$, the estimate of the location of the regression line as given in equation 4.14, for a particular value of x is given by:-

$$s_y^2 = s_{y.x}^2 \left[1/n + (x - \bar{x})^2 / \sum (x_i - \bar{x})^2 \right] \qquad (4.31)$$

so that the $100(1 - \alpha)\%$ confidence limits on the location of the regression line at x are given by:-

$$y \pm t_{\alpha/2} \, s_y \qquad (4.32)$$

where $t_{\alpha/2}$ is the value of Student's t with $n - 2$ d.f. obtained from Table A3 in the Appendix. Note that in equation 4.31 all the terms, except x, are constant for a particular regression, so that s_y^2 is a minimum when $x = \bar{x}$, and its value increases symmetrically as x deviates from $\bar{x}$. This gives confidence limits to the regression line which plot as curves as shown in Fig. 4.5 and simply reflects the errors in the estimation of the slope and intercept of the line. Obviously, if there is error in the slope of the line, the further one gets from the mean, the greater the error in y.

It is important to distinguish between these confidence limits and those of the next section. In this section the confidence limits apply to the location of the regression line at a particular value of x, i.e. they are the mean value of the y's that would be obtained by repeated random sampling at the particular value of x. In the next section the confidence limits are applicable to the single value of y obtained when a single random sample, with a specific value of x, is drawn from the population.

Numerical example

Assuming the data of Table 4.1 to be a random sample, we will calculate the 95% confidence limits on our estimates of the depth to the Benioff zone (y) for K% (x) = 1.5935 (its mean value), which gives $\hat{y} = 177$, and for x = 3.0, which gives $\hat{y} = 204$. From Table 4.2 we find that:-

$$s_{y.x}^2 = 236.0929$$

so that using equation 4.31 we may obtain s_y^2 from the expression:-

$$s_y^2 = 236.0929 \left[1/17 + (x - 1.5935)^2 / 8.7546 \right]$$

When x = 1.5935 this becomes:-

$$s_y^2 = 236.0929 \, (0.058824 + 0.0) = 13.8878 \quad \text{and} \quad s_y = 3.727$$

Table A3 in the Appendix gives $t_{0.025} = 2.131$ with 15 d.f. so that using equation 4.32 we obtain the confidence limits of the depth to the underlying

Benioff zone as 177 ± 2.131 x 3.727 which gives 177 ± 8. However, when x = 3.0 we have:-

$$s_y{}^2 = 236.0929 \ (0.058824 + 1.4065^2 \ / \ 8.7546) = 67.2369 \quad \text{and} \quad s_y = 8.200$$

which gives confidence limits for the depth to the Benioff zone of 204 ± 2.131 x 8.2021 or 204 ± 18. These 95% confidence limits are also shown in Fig. 4.5.

4.3.4 Confidence limits for an individual y value

If $\hat{y}$ is an estimate of an individual y, as opposed to the regression line value, then the variance of $\hat{y}$ is given by (e.g. Draper & Smith, 1966, p.24):-

$$s_y{}^2 = s_{y.x}{}^2 \left[1 + 1/n + (x - \bar{x})^2 \ / \ \sum (x_i - \bar{x})^2 \right] \qquad (4.33)$$

which is similar to equation 4.31 except for the additional term of unity within the brackets. The confidence limits can then be calculated using equation 4.32 as before.

Numerical example

We will continue with the data of the previous example, but will now use equation 4.33 to set the confidence limits on an individual value of y. When x = 1.5935 we have:-

$$s_y{}^2 = 236.0929 \ (1.0 + 0.058824) = 249.9807 \quad \text{and} \quad s_y = 15.811$$

so that the 95% confidence limits on an individual value of y at x = 1.5935 are given by 177 ± 2.131 x 15.811 or 177 ± 34, which are much wider than those in the previous example. Similarly, when x = 3.0 we obtain:-

$$s_y{}^2 = 236.0929 \ (1.0 + 0.058824 + 0.225966) = 303.3298 \quad \text{and} \quad s_y = 17.416$$

so that the confidence limits on an individual value of y at x = 3.0 are given by 204 ± 2.131 x 17.416 or 204 ± 37. These 95% confidence limits are also shown in Fig. 4.5.

To make the meaning of these confidence limits perfectly clear, if data from another randomly selected volcano in Java indicates that K% is 3.0 at Si% = 51, then one can be 95% confident that the Benioff zone lies between 204 ± 37km. (i.e. between 167km. and 241km.) beneath that particular volcano.

4.4 FUNCTIONAL AND STRUCTURAL RELATIONSHIPS

In the example of linear regression, the choice of independent and dependent variable was fairly logical - by observing the K% at Si% = 51 (the independent variable) predictions could be made about the depth to the underlying Benioff zone (the dependent variable). However, in many instances the choice is not obvious. For example, both variables may be subject to error, or one may be more interested in the slope of the line than in its predictive properties, which is often the case with age-dating methods and with variation diagrams. Under these circumstances the methods of either functional or structural relationships should be used, and excellent discussions of these two methods are given by Mark & Church (1977) and Jones (1979). This is basically the approach used by geochronologists for the interpretation of isochrons where the slope of the line is important e.g. Brooks et al. (1968, 1972), McIntyre et al. (1966) and York (1969). Although there are subtle differences in the assumptions of the two methods, they both give the same answers numerically.

It can be shown (Lindley, 1947) that when both variables are subject to error, b_f, the slope of the "best-fit" line:-

$$y = a_f + b_f x \tag{4.34}$$

is given by the general expression:-

$$b_f = \left[SS_y - \lambda SS_x + \sqrt{\left((SS_y - SS_x)^2 + 4\lambda SP_{xy} \right)} \right] / 2SP_{xy} \tag{4.35}$$

where λ is the ratio of the variances of the random errors in y and x. If replicate observations have been made it is possible to determine λ from the data, but in general λ has to be estimated or assumed. Some suggestions for doing this are reviewed by Jones (1979). Of particular interest are the results obtained with four specific values of λ :-

1. When $\lambda = 0$ (y is known without error), $b_f = SS_y/SP_{xy} = 1/b'$, where b' is the regression coefficient for the regression of X on Y i.e. $x = a' + b'y$.
2. When λ is infinite (x is known without error), $b_f = SP_{xy}/SS_x = b$, the regression coefficient for the regression of Y on X as given in equation 4.14.
3. When $\lambda = 1$, the solution to equation 4.34 gives b_p, which is known as either "Pearson's major axis solution" or the "principal axis solution". This is the same result as that given by the 1st eigenvector of principal components analysis (see Chapter 7 for further details).
4. When $\lambda = SS_y/SS_x$, $b_f = \sqrt{(SS_y/SS_x)} = b_m$, with b_m taking the same sign as r, the correlation coefficient. This is called the "reduced major axis solution" and is the method favoured by many authors.

As λ varies from 0 to infinity it can be shown that b_f varies continuously from $1/b'$ to b. It therefore follows that:-

$$b \leqslant b_f \leqslant b / r^2 \tag{4.36}$$

The geometric interpretation of the criteria that are used to fit the line for each of the four special cases mentioned above, are shown diagramatically in Fig. 4.6.

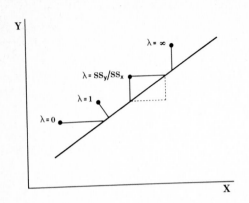

Fig. 4.6. Diagram to illustrate how deviations from the fitted lines are minimized for the four specific values of λ used in equation 4.31. When $\lambda = 0$, which is the regression of X on Y, the distances parallel to the x axis are minimized; when $\lambda = 1$, which is Pearson's major axis or the principal axis solution, the perpendicular distances are minimized; when $\lambda = SS_y/SS_x$, which is the reduced major axis solution, the rectangular areas are minimized; and when λ is infinite, which is the regression of Y on X, the distances parallel to the y axis are minimized.

Numerical example

We will again use the data of Table 4.1 to illustrate the differences between the "best-fit" lines for the four specific solutions to equation 4.35:-

Case 1. $\lambda = 0$, $b_f = 1/b'$, regression of X on Y
$b_f = SS_y / SP_{xy} = 6988.2353 / 173.7118 = 40.23$
$a_f = 176.4706 - 40.2289 \times 1.5935 = 112.4$

Case 2. λ is infinite, $b_f = b$, regression of Y on X
$b_f = SP_{xy} / SS_x = 173.7118 / 8.7546 = 19.84$
$a_f = 176.4704 - 19.8423 \times 1.5935 = 144.9$

Case 3. $\lambda = 1$, $b_f = b_p$, Pearson's major axis or principal axis solution
$SS_y - SS_x = 6988.2353 - 8.7546 = 6979.4807$
$b_f = \left[6979.4807 + \sqrt{(6979.4807^2 + 4 \times 173.7118)}\right] / (2 \times 173.7118) = 40.18$
$a_f = 176.4706 - 40.1786 \times 1.5935 = 112.4$

Case 4. $\lambda = SS_y / SS_x$, $b_f = b_m$, reduced major axis solution
$b_f = \sqrt{(SS_y / SS_x)} = \sqrt{(6988.2353 / 8.7546)} = 28.25$
$a_f = 176.4706 - 28.2531 \times 1.5935 = 131.4$

These four lines, together with the original data from Table 4.1 are plotted in Fig. 4.7.

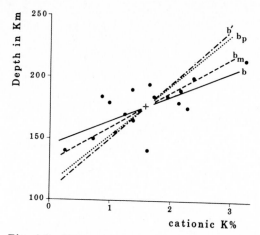

Fig. 4.7. Diagram to illustrate the lines obtained when using the four specific values of λ in equation 4.35 with the depth to the Benioff zone versus K% data of Table 4.1. Line b is the regression of Y on X, when λ is infinite. Line b_m is the reduced major axis solution, when $\lambda = SS_y/SS_x$. Line b_p is Pearson's major axis or the principal axis solution, when $\lambda = 1$. Line b' is the regression of X on Y, when $\lambda = 0$.

4.4.1 Significance tests

Significance tests for functional and structural relationships are not as well developed as they are for regression analysis. Creasy (1956) showed that for the special case when $\lambda = 1$, i.e. the principal axis solution, the test of the hypothesis that the slope of the regression line is zero, is the same as the null test for the correlation coefficient given in equation 4.7. However, the test was expressed in terms of φ (Greek letter phi), the angle of the line from the horizontal, so that confidence limits might be applied when $\varphi \neq 0$ by rotating the axes, but as pointed out, certain ambiguities arise in the interpretation of the results. Moran (1971) discusses the problem further and concludes that if no information is available concerning the value of λ, all that can be done is to calculate the confidence limits for the two regression lines as in equation 4.28 and to "assert that the true relationship lies in the region bounded by them". Jones (1979) also discusses this problem.

CHAPTER 5

THE GEOMETRY OF PETROLOGICAL DATA SPACE

5.1 INTRODUCTION

At this stage it is necessary to introduce the concept of p-dimensional geometry, in order to facilitate the understanding of some of the concepts introduced later in the book, in particular the petrological mixing models discussed in the next chapter.

Just as a point in 3-dimensional space may be defined by 3 coordinates along 3 axes, so a point in p-dimensional space may be defined by p coordinates along p axes. A single chemical analysis can, therefore, be thought of as a single point in a p-dimensional space specified by p oxide axes. Each oxide value in the analysis then becomes the coordinate of the point along the respective oxide axis and a series of n analyses become n points in p-dimensional space. The Harker diagram can then be thought of as $(p - 1)$ projections of the data from the p-dimensional space into specific 2-dimensional spaces.

In order to comprehend the relationship between analyses in p-dimensional space it is necessary to understand some of the principles of p-dimensional coordinate geometry. It is ironical that, although most petrologists have a good understanding of 3-dimensional geometry and its mathematics from structural geology and crystallography, few have extended this knowledge into p-dimensions to deal with their chemical data (see however, Greenwood, 1967; Le Maitre, 1968).

One of the niceties of p-dimensional space is that in many cases p can be thought of as 3, as most of the formulae of 3-dimensional coordinate geometry extend logically into p-dimensions. However, there are some differences involved in the geometric interpretation of 3-dimensional space and p-dimensional space that must be clearly understood in order to avoid any misinterpretations. For a detailed explanation of the geometry of p dimensions the reader is referred to some of the classic texts e.g. Sommerville (1929).

The nomenclature of geometric regions in p-dimensional space is summarized in Table 5.1 and is a logical extension of the terms used in 3-dimensional space. The generalized terms 1-flat, 2-flat, etc, define the number of degrees of freedom of movement within the particular region, e.g. in a plane or 2-flat there are 2 degrees of freedom. The term hyperplane is an

TABLE 5.1

Summary of the nomenclature of geometric regions in p-dimensional space.

Number of linearly independent points required to define geometric region	Generalized name	Specific name of geometric region in p-dimensional space
1	0-flat	point
2	1-flat	line
3	2-flat	plane
4	3-flat	hyperplane
.	..	..
p	(p-1)-flat	hyperplane
p+1	p-flat	polytope

alternative term for all k-flats where $4 \leqslant k \leqslant p$. A polytope is the p-dimensional space equivalent to a polyhedron in 3-dimensional space and a polygon in 2-dimensional space.

From Table 5.1 and knowledge of 3-dimensional space, it should be obvious that (Sommerville, 1929):-

an arbitrary r-flat and s-flat taken together will lie
in the same (r + s + 1)-flat

$$(5.1)$$

This relationship holds true as long as $r + s + 1$ is not greater than p, where p is the number of dimensions. For example, a point (0-flat) and a line (1-flat) will lie in a plane (2-flat). Similarly a line and a plane will lie in the same 4-flat in p-dimensional space. However, if $r + s + 1$ is greater than p it means that the two regions will intersect each other. Sommerville (1929) generalizes this as follows:-

an arbitrary r-flat and s-flat, which are both contained in an p-flat,
will intersect in a (r + s - p)-flat, provided that $(r + s + 1) \geqslant p$.
If $(r + s) < p$ they will have no point in common.

$$(5.2)$$

For example, in 3-dimensional space a line (1-flat) and a plane (2-flat) will, in general, intersect in a point (1+2-3 = 0-flat), while two planes (2-flats) will, in general, intersect in a line (2+2-3 = 1-flat). However, in 9-dimensional space, two planes (2-flat) need not intersect as $r + s < p$. Instead, each of the two planes will have a *point of closest approach* to the other, in an analogous manner to two skew lines in 3-dimensional space. Similarly, in 9-dimensional space a 3-flat and a 4-flat will have points of closest approach to each other. An understanding of this type of relationship is of utmost importance in generalized petrological mixing models (section 6.3.4) when it comes to studying such problems as whether or not a particular igneous rock trend would intersect petrogeny's residua system or whether two metamorphic assemblages intersect, i.e. the "n-dimensional tie-line" problem of Greenwood (1967).

One restriction that applies to all major element analyses is that the sum of the oxides is approximately 100% - this is referred to as the closure problem and has already been discussed in section 4.2.3. Geometrically, however, this means that all the analyses in p-dimensional space will lie very close to the hyperplane which intersects each oxide axis at 100%. Analyses with totals less than 100% will lie on the side of the hyperplane nearest the origin, while those with totals greater than 100% will lie on the side away from the origin. Major element analyses, therefore, lie very close to a $(p - 1)$ hyperplane. This is illustrated in Fig. 5.1 with a simple 3-dimensional example. If the 3 variables, Or, Ab and An are recalculated to 100% and plotted into the 3-dimensional space defined by the 3 axes Or, Ab, An, all the points will then lie on the equilateral triangle defined by the plane which intersects the 3 axes at 100%. This, of course, is the familiar 2-dimensional triangular diagram, Or-Ab-An, so popular among petrologists.

5.2 DISTANCE BETWEEN TWO POINTS

In the simple 2-dimensional case, d, the distance between two points X_1 and X_2, whose coordinates are (x_{11}, x_{12}) and (x_{21}, x_{22}), respectively, is given by the Pythagoras theorem:-

$$d = \sqrt{\left[(x_{11} - x_{21})^2 + (x_{12} - x_{22})^2\right]} \tag{5.3}$$

In p-dimensional space this extends to:-

$$d = \sqrt{\left[(x_{11} - x_{21})^2 + (x_{12} - x_{22})^2 + \dots + (x_{1p} - x_{2p})^2\right]} \tag{5.4}$$

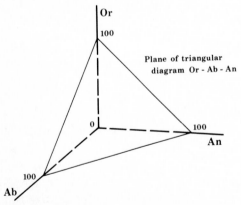

Fig. 5.1. Geometric interpretation of the closure effect. In the 3-dimensional space defined by Or, Ab and An, feldspar analyses expressed entirely in terms of Or, Ab and An will always lie in the 2-dimensional space defined by the equilateral triangle with its corners at 100% on each axis. This is the familiar triangular diagram Or-Ab-An.

or in more convenient notation:-

$$d = \sqrt{\sum (x_{1i} - x_{2i})^2}$$

(5.5)

These distances, which are often referred to as Euclidean distances, can be used in a purely geometric way to construct, for example, the shape of petrogeny's residua system Qz-Ne-Kp in p-dimensional space, as illustrated in Fig. 5.2 and in the worked example below.

Distance may also be used as a measure of similarity between two objects, as the closer they are the more similar they appear to be. It is often used this way in cluster analysis (see Chapter 10), although in this case the distance is often divided by the square root of the number of variables.

Numerical examples

  Case 1. Using the theoretical compositions of end-members in petrogeny's residua system, taken from Table A12 in the Appendix, the distance between Or and Ab in the weight percent space is given by:-

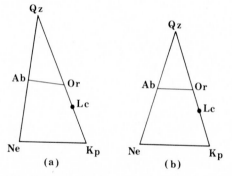

(a)   (b)

Fig. 5.2. The shape of petrogeny's residua system in (a) weight percent oxide space and (b) molecular percent and atomic percent space. Abbreviations are:-
Qz = quartz, Ab = albite, Ne = nepheline, Or = orthoclase, Lc = leucite, Kp = kaliophilite.

TABLE 5.2

Euclidean distances between the mineral molecules shown in Fig. 5.2, calculated using equation 5.4 with compositions from Table A12 in the Appendix. These distances were used to construct Fig. 5.2a.

Euclidean distances

	Qz	Ab	Ne	Or	Lc
Ab	39.66	Ab			
Ne	71.37	32.71	Ne		
Or	43.17	21.05	39.69	Or	
Lc	55.05	28.42	35.52	11.88	Lc
Kp	75.97	46.21	37.34	32.80	20.91

$$d^2 = (68.74 - 64.76)^2 + (19.44 - 18.32)^2 + (11.82 - 0.0)^2 + (0.0 - 16.92)^2$$
$$= 15.8404 + 1.2544 + 139.7124 + 286.2864 = 443.0936 \quad \text{or} \quad d = 21.05$$

This distance, together with all the other distances between the end-members are also given in Table 5.2 and are the data from which Fig. 5.2a was constructed. As can be seen, the shape of petrogeny's residua system in weight percent space is far from the equilateral triangle that is usually used. Fig. 5.2b shows the shape of the same system in molecular and atomic space and readers may care to check for themselves that the diagram is correct.

Case 2. An example of the use of distance as a similarity measure is given in Table 5.3. The problem was to find analyses that were most similar to the analysis of a lava from Katmai, Alaska (anal. 1). This was done with a program called MATC04, which is one of a series of programs in the CLAIR data system (see section 12.3.1). The program searched the CLAIR data file of 26373 igneous rock analyses and calculated the distance between each analysis and the one to be matched to find a specified number of nearest neighbours. Two distances were used. In the 9-dimensional space defined by SiO_2, TiO_2, Al_2O_3, Fe_2O_3, FeO, MgO, CaO, Na_2O and K_2O recalculated to 100%, 1046 analyses were found closer than 5.0 wt% units, and of those 190 were nearer than 2.5% and 3 were nearer than 1.0%. In the 3-dimensional space defined by CaO, Na_2O and K_2O, 568 analyses were found closer than 5.0 wt%, and of those 212 were nearer than 2.5% and 33 nearer than 1%.

Note that although analyses 2, 3 and 4 are three of the nearest analyses to analysis 1 in 9-dimensional space, they are not so close in the 3-dimensional space. Similarly, analyses 5, 6 and 7 are the three nearest in 3-dimensional space, yet two of them are a considerable distance from analysis 1 in 9-dimensional space, thus emphasizing one of the dangers of projections. The choice of which data space to use, of course, is entirely a matter for the individual petrologist and depends upon the problem to be solved.

5.3 DEFINITION OF DIRECTIONS IN SPACE

Consider a point P whose coordinates in 3-dimensional space are (x_1, x_2, x_3). These coordinates may be used to define the direction OP where O is the origin as illustrated in Fig. 5.3. However, this is not a very convenient way of defining the direction of OP, as the coordinates of any point along the line would also define the same direction. A more convenient way is to define the direction of OP in terms of the 3 direction angles, X_1OP, X_2OP and X_3OP, which are, of course, independent of the location of P.

Now from Fig. 5.3 it can be seen that v_1, the cosine of the angle X_1OP, is given by $v_1 = x_1 / OP$ but, using equation 5.5, $OP = \sqrt{(x_1^2 + x_2^2 + x_3^2)}$ so that we can write:-

$$v_1 = x_1 / \sqrt{(x_1^2 + x_2^2 + x_3^2)} \tag{5.6}$$

The values v_1, v_2 and v_3 defined in this manner are called *direction cosines*. It then follows that any set of coordinates $(x_1, x_2, ... x_p)$ in p-dimensional space, can be converted into direction cosine form by using the relationship that:-

$$v_i = x_i / \sqrt{\sum x_j^2} \tag{5.7}$$

from which it can be seen that:-

$$\sum v_i^2 = 1 \tag{5.8}$$

TABLE 5.3

An example of the use of distances as a similarity measure for matching analyses. The data were obtained using the CLAIR data system. See text for further details.

Analyses used	1	2	3	4	5	6	7
SiO_2	64.16	64.37	64.05	63.83	59.37	65.02	60.61
TiO_2	0.67	0.80	0.87	0.60	0.90	0.63	1.09
Al_2O_3	15.70	15.46	15.55	15.53	17.39	16.77	15.44
Fe_2O_3	3.80	3.81	4.21	3.80	2.81	4.21	2.10
FeO	2.04	1.80	1.70	3.01	3.74	2.61	6.78
MgO	2.55	2.63	2.74	2.60	31.0	1.15	2.80
CaO	5.53	5.67	5.25	5.59	6.34	4.81	5.57
Na_2O	3.67	3.47	3.53	3.24	4.23	3.21	3.73
K_2O	1.86	2.00	2.09	1.79	2.14	1.61	1.87

Distance from analysis 1 in 9-dimensional space							
	0	0.51	0.74	1.13	5.57	2.27	6.19

$CaO - Na_2O - K_2O$ ratios	1	2	3	4	5	6	7
CaO	50.00	50.90	48.30	52.64	49.88	49.95	49.87
Na_2O	33.18	31.15	32.47	30.51	33.28	33.33	33.39
K_2O	16.82	17.95	19.23	16.85	16.84	16.72	16.74

Distance from recalculated analysis 1 in 3-dimensional space							
	0	2.49	3.03	3.75	0.16	0.19	0.26

Key to analyses.
1. Lava boulder, Katmai, Alaska (Fenner, 1926) (8287)
2. Andesine hornblende pyroxene dacite, Milos, Aegean (Paraskevopoulos, 1956) (9438)
3. Dacite, Semangka Valley, S. Sumatra (Westerveld, 1952) (3854)
4. Andesite, Kizimen Volcano, Kamchatka (Vlodavetz & Piip, 1959) (4067)
5. Andesine augite dacitoide, Lesbos, Aegean (Paraskevopoulos, 1956) (9637)
6. Hornblende augite hypersthene trachyandesite, Aira Volcano, Japan (Ono, 1962) (5503)
7. Quartz diorite, Aulanko region, S.W. Finland (Simonen, 1948) (2424)

The number in brackets after each reference is the unique ISEQ number of the analysis in the CLAIR master data file.

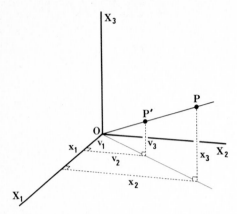

Fig. 5.3. Diagram showing the relationship between the direction cosines of a line and the coordinates of any point P (x_1, x_2, x_3) on the line. If OP' is of unit length then the coordinates of P' (v_1, v_2, v_3) are also the direction cosines of the line OP.

Now as equation 5.8 dictates that in Fig. 5.3, $v_1^2 + v_2^2 + v_3^2 = 1$, it also follows that the direction cosines are the coordinates of point P' on the line OP, where the distance OP' is 1 unit.

To determine the direction cosines of a line not passing through the origin, two points have to be defined on the line. Let these two points be X_1 and X_2 with coordinates $(x_{11}, x_{12}, ... x_{1p})$ and $(x_{21}, x_{22}, ... x_{2p})$, respectively. By translating the origin to point X_1, the new coordinates of point X_2 become $(x_{21} - x_{11}, x_{22} - x_{12}, ... x_{2p} - x_{1p})$. Using equation 5.7 it then follows that the direction cosines of the line are given by:-

$$v_i = (x_{2i} - x_{1i}) / \sqrt{\sum (x_{2j} - x_{1j})^2} \tag{5.9}$$

Finally, θ (Greek letter theta), the angle between two directions whose direction cosines are given by the vectors $\mathbf{V}_1 = (v_{11}, v_{12}, ... v_{1p})$ and $\mathbf{V}_2 = (v_{21}, v_{22}, ... v_{2p})$ is given by the simple relationship:-

$$\cos \theta = \sum v_{1i} v_{2i} \tag{5.10}$$

Multiplying any set of direction cosines by -1, only defines the opposite direction of the vector, since if $\mathbf{V}_1$ are a set of direction cosines, the angle θ between $\mathbf{V}_1$ and $-\mathbf{V}_1$ is given by:-

$$\cos \theta = \sum (-v_{1i} v_{1i})$$

$$= - \sum (v_{1i})^2 = -1 \quad \text{using equation 5.8}$$

therefore, $\theta = 180^\circ$.

It is important to understand direction cosines as many of the statistical procedures produce results in this form e.g. principal components analysis (Chapter 7) and multiple discriminant analysis (Chapter 9).

In addition, angular measurements can be used for many other purposes.

For example, the similarity between two igneous rock trends can be judged by measuring the angle between them. In another application Le Maitre (1968) used equations 5.9 and 5.10 to measure the angles between some linear functions of oxides commonly plotted in variation diagrams. These equations could also have been used to calculate the angles necessary to construct Fig. 5.2 rather than using distances.

The angle subtended by two points at the origin can also be used as a similarity measure, especially for data in which the important characteristics are the ratios between the variables. For example, suppose a rock is being altered and that the ratios of certain elements, which are characteristic of the original rock type, are remaining constant, while the absolute values of elements are being "diluted" by the alteration process. In terms of Fig. 5.3, this means that the coordinates are moving from point P towards O. It thus follows that, if the "dilutent" is not included as a variable, the angle calculated will indicate original similarity and be independent of the total amount of alteration. To use distance as a similarity measure, all the data would have to be recalculated on a "dilutent" free basis before calculating the distance. Such a method could also be useful with cumulative rocks.

This type of similarity measure is often used in palaeontology and biological sciences to eliminate the effect of size on similarity. It is also used in Q-mode factor analysis (section 8.3) and in cluster analysis (section 10.2.1).

Numerical examples

Case 1. To convert the analysis of Ab given in Table 5.2 into direction cosine form using equation 5.7:-

$$\sum x_i^2 = 68.74^2 + 19.44^2 + 11.82^2 + 0.0^2 = 5242.8136$$

$$v_1 = 68.74 / \sqrt{5242.8136} = 0.9494$$

Similarly, $v_2 = 0.2685$, $v_3 = 0.1632$ and $v_4 = 0.0$. As a check, note that to within rounding-off errors $v_1^2 + v_2^2 + v_3^2 + v_4^2 = 1.0$. Likewise, for Or, $v_1 = 0.9332$, $v_2 = 0.2640$, $v_3 = 0.0$, and $v_4 = 0.2438$.

Case 2. The angle subtended by Or and Ab at the origin is then given by equation 5.10:-

$$\cos \theta = (0.9494 \times 0.0332) + (0.2685 \times 0.2640) + 0.0 + 0.0 = 0.9569$$
and $\theta = 16.9^o$

The reader may care to check that the angle subtended by Lc and Kp at the origin is 19.1°, indicating that, according to the angular similarity measure, Or and Ab are more similar to each other than Lc and Kp. This is opposite to the results obtained using distance as a similarity measure (see Table 5.2) thus illustrating some of the problems of chosing similarity measures.

Case 3. Using equation 5.9 the direction cosines of the line joining Qz and Kp are calculated as follows:-

$\sum (x_{1i} - x_{2i})^2 = (100.0 - 37.99)^2 + (-32.23)^2 + 0.0 + (-29.78)^2 = 5770.8614$

therefore $v_1 = 0.8163$, $v_2 = -0.4243$, $v_3 = 0.0$ and $v_4 = -0.3920$. Note that exactly the same direction cosines would be obtained using Qz-Lc and Qz-Or as they all define the same line. Similarly, the direction cosines for the line Qz-Ne are $v_1 = 0.8085$, $v_2 = 0.5029$, $v_3 = -0.3057$ and $v_4 = 0.0$. Using equation 5.10, the angle Ne-Qz-Kp is:-

$\cos \theta = (0.8163 \times 0.8085) + (-0.4243 \times -0.5029) + 0.0 + 0.0 = 0.8734$
or $\theta = 29.2^{\circ}$

The reader may care to check that this is the angle in Fig. 5.2a.

Case 4. Another example of the use of angles as a similarity measure is shown in Table 5.4, where all possible angles between eight approximately linear igneous rock trends have been calculated. The direction cosines of the trends are taken from Le Maitre (1968, Table 2) by using the first eigenvector from principal components analysis which gives the "best straight line fit" in p-dimensional space (see section 6.2). As an exercise the reader may like to check some of the angles.

As might be expected, the angles indicate that the trends of the three oceanic alkali provinces, St. Helena, San Miguel and Mauritius, are fairly close to each other. Likewise, the angles for the three sub-alkali trends of Crater Lake, Katmai and Hachijo-jima indicate that their trend directions are close to each other. The Thingmuli trend does not seem to be very close to either the oceanic alkali or sub-alkali trends, while the strongly alkaline Highwood Mountains trend is markedly different from all the other trends. In fact, there is as much angular difference between the Highwood Mountains trend and the oceanic alkali trends as there is between the sub-alkali and oceanic alkali trends.

TABLE 5.4

An example of the use of angles as a measure for judging the similarity of the directions of igneous rock trends. The direction of the trend for each province, as defined by the 1st eigenvector, was obtained from Le Maitre (1968).

Calculated angles in degrees between trends, using equation 5.10

	St.Helena						
San Miguel	6.4	San Miguel					
Mauritius	8.5	7.7	Mauritius				
Highwood Mts.	29.3	31.0	25.3	Highwood Mts.			
Crater Lake	22.5	22.2	26.5	47.9	Crater Lake		
Katmai	26.8	26.2	31.6	53.0	6.8	Katmai	
Hachijo-jima	24.4	23.4	27.9	50.5	5.4	7.3	Hachijo-jima
Thingmuli	17.9	14.9	20.7	44.2	13.2	14.5	12.4

5.4 EQUATION OF A LINE AND THE ADDITION-SUBTRACTION PROBLEM

In two dimensions, the equation of a line passing through two points $X_1 = (x_{11}, x_{12})$ and $X_2 = (x_{21}, x_{22})$, can be written as a single equation:-

$$\frac{x_1 - x_{11}}{x_{21} - x_{11}} = \frac{x_2 - x_{12}}{x_{22} - x_{12}} \tag{5.11}$$

where $X = (x_1, x_2)$ is any point on the line. In p-dimensional space, however, the equation of a line passing through the two points $X_1 = (x_{11}, \dots, x_{1p})$ and $X_2 = (x_{21}, \dots, x_{2p})$ is given by the equation:-

$$\frac{x_1 - x_{11}}{x_{21} - x_{11}} = \frac{x_2 - x_{12}}{x_{22} - x_{12}} = \dots\dots\dots = \frac{x_p - x_{1p}}{x_{2p} - x_{1p}} \tag{5.12}$$

where $X = (x_1, x_2, \dots x_p)$ is any point on the line. If one of the x_i's is specified, the remaining p - 1 values are fixed and may be determined by substitution in equation 5.12. Specifying one value of x_i simply fixes the value of the ratios in the equations, so that it is often convenient to rewrite it to give the value of any x_i along the line, as:-

$$(x_i - x_{1i}) / (x_{2i} - x_{1i}) = d \tag{5.13}$$

or

$$x_i = d(x_{2i} - x_{1i}) + x_{1i} \tag{5.14}$$

where d is the distance from point X to X_1, and the unit of measure of d is the distance between X_1 and X_2. This can be seen by rewriting equation 5.14 as:-

$$x_i = x_{1i}(1 - d) + x_{2i}d \tag{5.15}$$

as when d = 0, $X = X_1$ and when d = 1, $X = X_2$. Written in this form, the mass balance relationship or "lever rule" between X, X_1 and X_2 can also be clearly seen, i.e.

$$1 \text{ part of } X = (1 - d) \text{ parts of } X_1 + d \text{ parts of } X_2 \tag{5.16}$$

This relationship holds true whatever the numerical value of d. For example:-

if d = 0.2 $X = 0.8X_1 + 0.2X_2$
if d = 2.0 $X = -X_1 + 2.0X_2$
if d = -2.0 $X = 3.0X_1 - 2.0X_2$

If both sides of equation 5.13 are multiplied by d_x, the distance between X_1 and X_2, we obtain:-

$$d_x (x_i - x_{1i}) / (x_{2i} - x_{1i}) = d_x d \tag{5.17}$$

But as $d_x = \sqrt{\sum (x_{2i} - x_{1i})^2}$ we can use the relationship of equation 5.9 to rewrite equation 5.17 as:-

$$(x_i - x_{1i}) / v_i = d' \tag{5.18}$$

or:-

$$x_i = d'v_i + x_{1i} \tag{5.19}$$

which is the equation of a line through a single point X_1 given $\mathbf{V} = (v_1, v_2, \ldots v_p)$, the direction cosines of the line. The new constant d' is then measured in the same units as the reference axes.

The principles involved in this section should be familiar to petrologists, as they are the mathematical basis of the well known addition-subtraction problem of Bowen (1928, p.75-77), i.e. what material should be added to or subtracted from one magma type to produce another. However, like Bowen, many petrologists still prefer to solve the equations graphically, or simply display the effect in variation diagrams with mineral "control lines", e.g. Cox & Jamieson (1974), Cox et al. (1979), Powers (1955). In problems involving two compositions, equation 5.14 should be used, i.e. the equations of a line through two points. There is, of course, no unique solution to any addition-subtraction problem as there are an infinite number of possible solutions (points on the line), so that it is up to the petrologist to argue for one position on the line rather than another, either from the point of view of the composition or the amount of material involved. In this type of addition-subtraction problem involving the relationship between two specific compositions, the amounts of materials to be added or subtracted are important.

Another type of addition-subtraction problem involves igneous rock trends. As many of these, or parts of them, are approximately straight lines (Le Maitre, 1968), it is often useful to be able to determine where the trend may have come from, or would go to, in p-dimensional space. This can be done by defining the location of the linear trend in p-dimensional space by means of an equation, in one of two ways:-

1. By defining two points on the trend, by averaging analyses from each end of the trend. Equation 5.14 can then be used to provide possible solutions, which will also include the proportion of the amounts of material to be added or subtracted. The problem has, therefore, been reduced to Bowen's addition-subtraction problem. This is the commonest and easiest approach, but not necessarily the best, and has been used many times in the literature, e.g. Baker (1968), Frey et al. (1978), Le Maitre (1962), Macdonald (1949), McBirney & Aoki (1968), although whether the solutions have been obtained mathematically or graphically is not always clear.

2. By defining one point on the trend and the direction of the trend. Equation 5.19 can then be used to calculate any composition on the trend. Defining the direction can be done by using principal components analysis (see Chapter 7) to obtain the first eigenvector, which can be regarded as the direction of the "best straight-line fit" in p-dimensional space; principal components analysis also gives the mean of all the data. This is a slightly

more elegant method than the previous one, as it uses all the data in the trend, but it will not provide any information concerning the amounts to be added or subtracted. In many instances, however, this is of no importance. A comparison of the two methods is given in the examples below.

These addition-subtraction problems, involving equations 5.14 or 5.19 are, of course, ideally suited for interactive computer use (Stormer and Nicholls, 1978). Having defined two compositions, or one composition and a set of direction cosines, it is a simple matter to enter a distance and to have the composition (and, if required, normative parameters) immediately printed out until a satisfactory solution is found.

TABLE 5.5

An illustration of the mathematical solution of the addition-subtraction problem that Bowen (1928, p.76) solved graphically to decide what would have to be removed or added to the Plateau magma type to produce the Non-porphyritic central magma type of Mull.

	Plateau magma type X_1	Non-porphyritic Central magma type X_2	Equations of line through X_1 and X_2 in form of equation 5.14	Value of d at which oxide is zero
SiO_2	45.0	50.0	5d + 45	-9.0
Al_2O_3	15.0	13.0	-2d + 15	7.5
$FeO+Fe_2O_3$	13.0	13.0	0d + 13	infinity
MgO	8.0	5.0	-3d + 8	2.66
CaO	9.0	10.0	d + 9	-9.0
Na_2O	2.5	2.8	0.3d + 2.5	-8.33
K_2O	0.5	1.2	0.7d + 0.5	-0.7143

	Solution A Material to be added to X_1 to produce X_2		Solution B Material to be subtracted from X_1 to produce X_2	
	Bowen's graphical solution	Mathematical solution with d=2.66	Bowen's graphical solution	Mathematical solution with d=-0.7143
SiO_2	58.0	58.3	41.5	41.4
Al_2O_3	9.0	9.7	16.0	16.4
$FeO+Fe_2O_3$	13.0	13.0	13.0	13.0
MgO	0.0	0.0	10.0 *	10.1
CaO	12.0	11.7	8.5 *	8.3
Na_2O	3.5	3.3	2.25	2.3
K_2O	2.0	2.4	0.0	0.0

* In Table VI of Bowen (1928, p.77) the values of MgO and CaO for solution S are interchanged. The graphical solution of Bowen's Fig. 22 is, however, correct.

Numerical examples

Case 1. Bowen (1928) posed the problem of what material must be added to or subtracted from the Plateau magma type of Mull (X_1, Table 5.5) to produce the Non-Porphyritic Central magma type (X_2, Table 5.5). He solved the problem graphically and chose as solutions the two limiting cases furthest from the two analyses where all the oxides have either positive values or are zero. These two solutions are also given in Table 5.5, together with the mathematical solution for the same two limiting cases. This is done by substituting the two compositions into equations 5.14 in order to calculate the value of d at which each oxide becomes zero. The values of d corresponding to the two limiting solutions are then given by the positive value with the smallest magnitude (i.e. d = 2.66 when MgO = 0) and the negative value with the smallest magnitude (i.e. d = -0.7143 when K_2O = 0). Substituting these two values of d into the set of equations then gives the required solutions, as shown in Table 5.5.

Using equation 5.16 the proportions of the material to be added (A) or subtracted (S) can be calculated as follows:-

$$A = -1.66X_1 + 2.66X_2$$

which can be rewritten as:-

$$100X_1 + 60.2A = 160.2X_2 \qquad \text{(Bowen gives 62.5\% A)}$$

and

$$S = 1.7143X_1 - 0.7143X_2$$

which can be rewritten as:-

$$100X_1 - 58.35S = 41.7X_2 \qquad \text{(not given by Bowen)}$$

A popular misconception is that the answer to any addition-subtraction problem must lie close to, or at least between, the two limiting solutions where one of the oxides becomes zero. However, in some processes this constraint may be misleading, as it is possible that some constituents are being removed while others are being added at the same time e.g. some metasomatic and alteration processes. Thus removing a composition which contains some negative oxides would simply mean that the negative oxides are being added instead of being removed. Also, as the next example will show, the limiting solutions are sometimes far from realistic.

Using the linear equations 5.14 and 5.19 by themselves is only one approach to the addition-subtraction problem. A more sophisticated approach is given in the next chapter, dealing with petrological mixing problems, although in most cases it is only a question of interpreting the single composition produced by equations 5.14 or 5.19 in terms of several mineral species.

Case 2. To illustrate the two methods of solving addition-subtraction problems involving igneous rock series, 17 analyses of volcanic rocks were chosen from Adak and Kanaga Islands in the Aleutian Islands (Coats, 1952). This series has been shown to be approximately linear and is group 27 of Le Maitre (1968, Fig. 2). Before being used the nine major oxides SiO_2, TiO_2, Al_2O_3, Fe_2O_3, FeO, MgO, CaO, Na_2O and K_2O of each analysis were recalculated to sum to 100%.

The 1st method involves calculating average compositions for each end of the trend, and using equation 5.14 to calculate any composition on the trend. To represent the more basic end of the trend five analyses with SiO_2 values between 50% and 55% were averaged. The acidic end of the trend is represented by another 5 analyses with SiO_2 values between 59% and 62%. These two compositions are given in Table 5.6 as B (Basic) and A (Acidic), respectively, together with the two limiting solutions B1 and A1, obtained by using equation 5.14. The direction cosines of the line between compositions A and B are given as VAB in Table 5.6.

The 2nd method involves defining the trend in terms of its direction cosines and its mean, and then using equation 5.19 to calculate any composition on the trend. This was done by using principal components analysis on the 17 analyses to obtain the mean of the data and the first eigenvector extracted from the variance-covariance matrix, to represent the direction of the trend. These values are given in Table 5.6 as M (Mean) and V (eigenVector), respectively, together with two other limiting solutions B2 and A2, obtained using equation 5.19.

As might be expected, there are slight differences between the two methods and the angle between the trends VAB and V is 5.4 (Table 5.7). The two acidic solutions A1 and A2 are fairly similar, being only a distance of 1.9% apart in 9-dimensional space, but the two basic solutions B1 and B2 are not in such good agreement being 6.5% apart. This is due to the fact that the basic solutions are at a much greater distance from the mean of the trend than the acidic solutions.

To give some general idea of the magnitude of the differences between the two methods, data from three other volcanic provinces, also used by Le Maitre (1968), were compiled and are given in Table 5.7, together with the data from Adak and Kanaga. The two sets of solutions for Crater Lake and Katmai are obviously in good agreement with each other, either due to the linearity of the trends or possibly due to the lucky choice of the analyses to be included in the two averages A and B. For Adak and Kanaga and Hachijo-jima, on the other hand, the agreement is not so good. In the cases of Crater Lake, Katmai and Hachijo-jima the limiting solutions all behave in the same manner as Bowen's classical example, with MgO and K_2O being zero in the acidic and basic solutions, respectively. Adak and Kanaga, on the other hand, is different as MgO goes to zero only in one of the acidic solutions. In the other, which should be more reliable as it is by method 2, FeO goes to zero. The basic solution for Adak and Kanaga, however, is most unusual as it is SiO_2 that has become zero, marginally ahead of K_2O! This is an excellent example of the dangers of assuming that one of the limiting solutions will give reasonable solutions.

Case 3. It is always wise to inspect a range of compositions around an area of interest, as it is often easier to detect unreasonable solutions and thus bracket a reasonable solution. The previous numerical example, where the

TABLE 5.6

A comparison of two possible methods of solving addition-subtraction problems with igneous rock trends, using volcanic data from Adak and Kanaga, Aleutian Islands (Coats, 1952).

METHOD 1 using two average compositions and equations 5.14

	B	A	B1	A1	VAB
SiO_2	52.01	60.47	0.0	73.3	0.901
TiO_2	0.72	0.50	2.1	0.2	−0.023
Al_2O_3	18.88	16.99	30.5	14.1	−0.201
Fe_2O_3	4.80	4.24	8.2	3.4	−0.060
FeO	4.50	2.73	15.4	0.1	−0.188
MgO	4.45	2.68	15.3	0.0	−0.188
CaO	9.64	7.11	25.2	3.3	−0.269
Na_2O	3.54	3.67	2.7	3.9	0.014
K_2O	1.46	1.60	0.6	1.8	0.015
Distance d used in equations 5.14			−6.148	2.514	

B − average of 5 basic analyses with SiO_2 between 50% and 55%
A − average of 5 acidic analyses with SiO_2 between 59% and 62%
B1 − one limiting solution with $SiO_2 = 0.0$
A1 − the other limiting solution with $MgO = 0.0$
VAB − direction cosines of line joining A and B

METHOD 2 using the mean, the direction of the trend and equations 5.19

	M	V	B2	A2
SiO_2	56.06	0.887	0.0	72.1
TiO_2	0.69	−0.022	2.1	0.3
Al_2O_3	17.77	−0.138	26.5	15.3
Fe_2O_3	4.20	−0.044	7.0	3.4
FeO	4.10	−0.227	18.4	0.0
MgO	3.70	−0.187	15.5	0.3
CaO	8.44	−0.323	28.9	2.6
Na_2O	3.52	0.030	1.6	4.1
K_2O	1.52	0.024	0.0	2.0
Distance d' used in equations 5.19		−63.202	18.062	

M − mean of 17 analyses
V − 1st eigenvector of 17 analyses (Le Maitre, 1968)
B2 − one limiting solution with $SiO_2 = 0.0$
A2 − the other limiting solution with $FeO = 0.0$

TABLE 5.7

A summary of the discrepancies between the two methods of solving
addition-subtraction problems, as given in Table 5.6, using trends from four
volcanic provinces. The terms VAB, V, A1, A2, B1 and B2 have the same
meaning as in Table 5.6.

	Number of group in Le Maitre (1968)	Angle in degrees between VAB & V	Distance between A1 & A2	Distance between B1 & B2	Number of analyses used
Adak & Kanaga	27	5.4	1.9	6.5	17
Crater Lake	19	0.0	0.6	0.4	25
Katmai	24	1.7	1.0	0.7	27
Hachijo-jima	40	3.2	1.4	10.4	28

basic limiting solution contains no SiO_2 and is obviously petrologically
unreasonable, is a case in point.

The composition of some possible acidic and basic solutions and their CIPW
norms are given in Tables 5.8 and 5.9, respectively, for a range of distances
using the composition M and direction cosines V of Table 5.6. As can be seen,
the basic solutions become more and more undersaturated until at composition
B6 there is a final silica deficiency, i.e. there is not enough SiO_2 to satisfy all
the silicates in the norm. If a basic solution is being sought, one somewhere in
the region of compositions B2 to B4 would seem reasonable. In view of the
undersaturated nature of these compositions, one might suspect that a mixture
of amphibole (which is often normatively undersaturated) and plagioclase could
be subtracted from the basic lavas to produce the more acidic lavas. Coats
(1952) inferred qualitatively that the trend was produced by the removal of
olivine and pyroxene from the basic lavas, but the chemistry of the basic
solutions do not seem to support this view.

If required, such ideas could be tested further in detail by using the
petrological mixing models described in section 6.3.

5.5 TRANSFORMATION OF COORDINATES

It can be shown (e.g. Jöreskog et al., 1976, p.32) that if point X has
coordinates $(x_1, x_2, \ldots, x_p)$ with respect to an original set of p reference axes,
its coordinates with respect to a new set of q reference axes will be $(z_1, z_2, \ldots z_q)$ where:-

$$z_i = \sum_{j=1}^{p} v_{ij} x_j \qquad (5.20)$$

TABLE 5.8

A range of 8 acidic compositions to illustrate the types of materials that mathematically could be responsible for producing the trend of the volcanic rocks from Adak and Kanaga (Coats, 1952). The data were produced using equation 5.19 and the mean (M) and direction cosines (V) of Table 5.6. Distance is the value of d' in equation 5.19.

	A1	A2	A3	A4	A5	A6	A7	A8
SiO_2	72.08	70.64	69.20	67.76	66.31	64.87	63.43	61.99
TiO_2	0.29	0.33	0.36	0.40	0.44	0.47	0.51	0.54
Al_2O_3	15.28	15.50	15.73	15.95	16.17	16.40	16.62	16.85
Fe_2O_3	3.41	3.48	3.55	3.62	3.69	3.76	3.89	3.91
FeO	0.00	0.37	0.74	1.11	1.48	1.84	2.21	2.58
MgO	0.32	0.63	0.93	1.23	1.54	1.84	2.15	2.45
CaO	2.61	3.13	3.66	4.18	4.71	5.23	5.76	6.28
Na_2O	4.06	4.01	3.96	3.92	3.87	3.82	3.77	3.72
K_2O	1.95	1.91	1.88	1.84	1.80	1.76	1.72	1.68
Distance	18.06	16.44	14.81	13.19	11.56	9.94	8.31	6.68

Weight percent CIPW norms

	A1	A2	A3	A4	A5	A6	A7	A8
Q	34.91	32.33	29.74	27.20	24.97	22.74	20.49	18.00
C	1.74	1.13	0.53	0.00	0.00	0.00	0.00	0.00
Or	11.54	11.31	11.08	10.85	10.62	10.39	10.16	9.93
Ab	34.37	33.96	33.55	33.13	32.72	32.31	31.89	31.48
An	12.93	15.53	18.14	20.52	21.47	22.41	23.36	24.31
Di	0.00	0.00	0.00	0.17	1.46	2.75	4.05	5.39
Hy	0.80	1.56	2.32	2.99	3.15	3.31	3.53	4.19
Hm	3.41	3.31	2.64	1.96	1.28	0.60	0.00	0.00
Mt	0.00	0.24	1.32	2.41	3.49	4.58	5.56	5.66
Il	0.00	0.62	0.69	0.76	0.83	0.90	0.96	1.03
Ru	0.29	0.00	0.00	0.00	0.00	0.00	0.00	0.00
Fe/Mg+Fe	0.00	0.00	0.00	0.00	0.00	0.00	0.84	7.17

and $V_i = (v_{i1}, \ldots v_{ip})$ are the direction cosines of the ith new reference axis, defined in terms of the original set of axes. The new axes, of course, do not have to be at right angles to each other.

In terms of geometry this is equivalent to projecting the point X onto the new reference axis V_i, as the value z_i is the distance of the point X from the hyperplane which passes through the origin of the reference axes and is normal to the direction whose direction cosines are V_i. This is illustrated in a simple 2-dimensional case in Fig. 5.4. If OX_1 and OX_2 are the original set of reference axes, OV_1 and OV_2 are the new set of reference axes, with direction cosines V_1 and V_2, respectively, and P is any point with coordinates (x_1, x_2),

TABLE 5.9

A range of 8 basic compositions to illustrate the types of materials that mathematically could be responsible for producing the trend of the volcanic rocks from Adak and Kanaga (Coats, 1952). The data were produced using equation 5.19 and the mean (M) and direction cosines (V) of Table 5.6. Distance is the value of d' in equation 5.19. Si.Def. = Final silica deficiency in the CIPW norm.

	B1	B2	B3	B4	B5	B6	B7	B8
SiO_2	50.46	47.57	44.69	41.81	38.92	36.04	33.16	30.27
TiO_2	0.83	0.90	0.97	1.04	1.12	1.19	1.26	1.33
Al_2O_3	18.64	19.09	19.54	19.99	20.44	20.88	21.33	21.78
Fe_2O_3	4.48	4.62	4.76	4.91	5.05	5.19	5.34	5.48
FeO	5.53	6.27	7.01	7.75	8.49	9.22	9.96	10.70
MgO	4.88	5.49	6.10	6.70	7.31	7.92	8.53	9.14
CaO	10.48	11.53	12.58	13.63	14.68	15.73	16.78	17.83
Na_2O	3.33	3.23	3.14	3.04	2.94	2.84	2.75	2.65
K_2O	1.37	1.29	1.21	1.13	1.06	0.98	0.90	0.82
Distance	-6.32	-9.57	-12.82	-16.07	-19.32	-22.57	-25.82	-29.07

Weight percent CIPW Norms

	B1	B2	B3	B4	B5	B6	B7	B8
Or	8.09	7.63	7.16	0.65	0.00	0.00	0.00	0.00
Ab	28.18	18.95	8.05	0.00	0.00	0.00	0.00	0.00
An	31.87	33.77	35.66	37.55	39.44	41.33	43.23	45.12
Ne	0.00	4.56	10.01	13.93	13.48	13.03	12.58	12.14
Lc	0.00	0.00	0.00	4.75	4.89	0.00	0.00	0.00
Ks	0.00	0.00	0.00	0.00	0.00	3.29	3.02	2.76
Di	16.21	18.92	21.64	24.35	8.47	0.00	0.00	0.00
Hy	2.74	0.00	0.00	0.00	0.00	0.00	0.00	0.00
Ol	4.83	7.77	8.72	9.68	17.17	22.08	24.00	25.93
Cs	0.00	0.00	0.00	0.00	7.10	11.36	12.39	13.41
Mt	6.49	6.70	6.91	7.11	7.32	7.53	7.74	7.94
Il	1.57	1.71	1.85	1.98	2.12	2.25	2.39	2.53
Si.Def.	0.00	0.00	0.00	0.00	0.00	0.87	5.35	9.83
Fe/Mg+Fe	24.17	25.69	26.86	27.80	28.56	29.19	29.72	30.18

equation 5.20 then gives z_1 and z_2, which are OP_1 and OP_2, respectively.

The main application of equation 5.20 is to enable data to be projected onto new axes, such as those defined by principal components analysis (Chapter 7) or multiple discriminant analysis (Chapter 9). It could also be used to rank data within an approximately linear igneous rock series by projecting each analysis onto the direction of the first eigenvector in p-dimensional space.

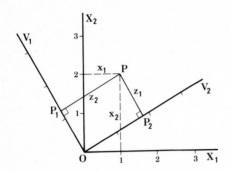

Fig. 5.4. Illustration of the transformation of coordinates or the projection of a point onto a new set of axes, using equation 5.20. Point P has coordinates (x_1, x_2) with respect to the original axes X_1 and X_2, and coordinates (z_1, z_2) with respect to the new set of axes V_1 and V_2.

Numerical example

In Fig. 5.4, angle X_2OV_1 is 30°. The direction cosines of OV_1 and OV_2 are, therefore, $(-0.500, 0.866)$ and $(0.866, 0.500)$, respectively. If P has coordinates $(1.0, 2.0)$, we can use equation 5.20 to obtain the new coordinates as:-

$$OP_1 = (1 \times -0.500) + (2 \times 0.866) = 1.232$$

and

$$OP_2 = (1 \times 0.866) - (2 \times 0.500) = 1.866$$

which can be checked with the scale in Fig. 5.4.

5.6 EQUATION OF A HYPERPLANE

By analogy with equation 5.15, the equation of a hyperplane or $(q - 1)$-flat passing through q independent points is given by a series of equations which can be written in matrix form as:-

$$\mathbf{X} = a_1\mathbf{x}_1 + a_2\mathbf{x}_2 + \ldots + a_q\mathbf{x}_q \tag{5.21}$$

where the vector $\mathbf{X}$ defines the coordinates of any point X on the $(q - 1)$-flat, vectors $\mathbf{x}_1, \mathbf{x}_2, \ldots \mathbf{x}_q$ define the coordinates of the q points $X_1, X_2, \ldots X_q$, and $a_1, a_2, \ldots a_q$ are constants which are subject to the condition that:-

$$\sum a_i = 1 \tag{5.22}$$

If $q = p$ equation 5.21 simplifies to a single equation:-

$$\sum v_i x_i = d \tag{5.23}$$

where $(x_1, x_2, \ldots x_p)$ are the coordinates of any point X in the $(p - 1)$-flat or hyperplane, $(v_1, v_2, \ldots v_p)$ are the direction cosines of the single normal to the hyperplane, and d is the perpendicular distance of the hyperplane from the origin.

If required, equation 5.10 could be used to calculate the angle between two (p - 1)-flats in p-dimensional space, as the angle between two planes is the same as the angle between their normals. However, by far the most important application in petrology is in modelling igneous rock trends and metamorphic reactions, as can be seen in the next chapter dealing with petrological mixing problems.

Numerical example

We will determine the equation of the 2-flat defined by the Or-Ab-An triangle in Fig. 5.1. Because of symmetry, the direction cosines of the normal to the plane must all be equal and, therefore, must be $1/\sqrt{3}$, as the sum of their squares must also be 1 (equation 5.8). Simple trigonometry also indicates that the perpendicular distance of the plane from the origin is $100/\sqrt{3}$. Using equation 5.23 the equation of the plane Or-Ab-An is therefore:-

$$Or/\sqrt{3} + Ab/\sqrt{3} + An/\sqrt{3} = 100/\sqrt{3}$$

which simplifies to:-

$$Or + Ab + An = 100$$

which, of course, should be obvious.

5.7 SOME LIMITATIONS ON THE USE OF RATIOS

The use of ratios in the construction of petrological variation diagrams is extremely common and ranges from the familiar triangular diagram to simple rectangular plots, frequently employed by geochemists. In general, the two ratios z_1 and z_2 can be expressed as:-

$$z_1 = b_1 x_p / x_q$$

and

$$z_2 = b_2 x_r / x_s \tag{5.24}$$

where x_p, x_q, x_r and x_s are four original variables, which are usually major or minor oxides or elements, normative components, linear functions of variables etc., and b_1 and b_2 are constants. Note that in this section, *p is not the number of variables.*

Plots constructed using the ratios of equations 5.24 suffer from one major disadvantage that makes graphical interpretation extremely difficult; namely that, *in general, a series of points which are colinear in the original space,* such as mixtures of two compositions, *will not be colinear in the ratio space,* but will lie on a curve, which may be either convex up or down. By writing expressions for the two ratios in terms of a proportion c of one composition and $(1 - c)$ of a second composition, and then eliminating the proportionality c, it can be shown that the general form of these curves in the ratio space is given by an equation of the type:-

$$a_1 z_1 + a_2 z_2 + a_3 z_1 z_2 + a_4 = 0 \qquad (5.25)$$

where a_1, a_2, a_3 and a_4 are constants. There are, however, some specific directions in the original space which will project as lines in the ratio plot. These are defined by the condition that makes a_3, in equation 5.25, equal to zero, which is that:-

$$x_{1q} x_{2s} = x_{2q} x_{1s} \qquad (5.26)$$

where the subscripts 1 and 2 refer to two compositions lying on the linear trend. These principles are illustrated in Fig. 5.5 using the data of Table 5.10.

Plots constructed with ratios in which the two denominators are the same variable (i.e. $q = s$ in equation 5.24), preserve linearity as condition 5.26 is always satisfied. These are the type of ratio advocated by Beswick & Soucie (1978) and Pearce (1968, 1970) for general use in variation diagrams.

Triangular diagrams can be considered as a special case of equation 5.24 in which $b_1 = b_2 = 100$, and the two denominators are the same, and equal to the sum of x_p, x_r and one other component. This can be expressed as:-

$$z_1 = 100 x_p / (x_p + x_r + x_t) \quad \text{and} \quad z_2 = 100 x_r / (x_p + x_r + x_t) \qquad (5.27)$$

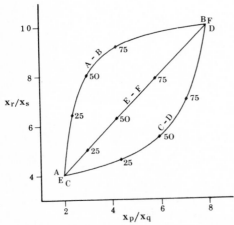

Fig. 5.5. Ratio plot using the data of Table 5.10, showing three loci for mixtures of the pairs of compositions A & B, C & D and E & F. The straight line in the original space between pair E & F, plots as a straight line in the ratio plot, as equation 5.26 is satisfied, but the straight lines in the original space between pairs A & B and C & D, plot as curves. The points on each curve where the mixtures represent 25%, 50% and 75% are also marked to illustrate that even when the projection is linear, the lever rule is not necessarily applicable.

TABLE 5.10

Data used to construct Fig. 5.5 and to illustrate some of the problems of the interpretation of ratio diagrams. The three compositions A, C and E, were all chosen to have identical ratios of the variables x_p/x_q and x_r/x_s, as were compositions B, D and F. The loci of mixtures of compositions A & B, C & D, and E & F, which are straight lines in the original variable space, project into the ratio space as curves for pairs A & B and C & D, and as a line for the pair E & F. The compositions AB, CD and EF are 50:50 mixtures of A & B, C & D, and E & F, respectively.

Variables	Compositions								
	A	AB	B	C	CD	D	E	EF	F
x_p	30	27	24	100	450	800	20	34	48
x_q	15	9	3	50	75	100	10	8	6
x_r	40	120	200	120	110	100	200	250	300
x_s	10	15	20	30	20	10	50	40	30
x_p/x_q	2	3	8	2	6	8	2	4.25	8
x_r/x_s	4	8	10	4	5.5	10	4	6.25	10

where x_t is the other or third component. Such triangular projections will always be linear, as equation 5.26 is always satisfied. The fact that, in conventional triangular diagrams, the z_1 and z_2 axes are plotted at 60 to each other does not, of course, effect the linearity.

However, even if linearity is preserved in the projection, all the ratio diagrams have the added limitation that, *in general, the lever rule is not applicable* (equation 5.16). This means that a mixture of equal amounts of two compositions will not plot halfway between the two compositions in the ratio plot – theoretically it can lie anywhere in between. It can also be shown that the lever rule is only applicable if, to use the notation of equations 5.26, $x_{1q} = x_{1s} = x_{2q} = x_{2s}$, which is the case for any triangular diagrams in which the sum of the three components is 100% (or any other constant). This means that the lever rule is applicable, for example, to a pure three component experimental phase diagram, but not to a triangular diagram into which the normative amounts of Or, Ab and An of a series of rock analyses have been projected as, in general, the sum of Or + Ab + An will be different for each analysis.

CHAPTER 6

MULTIPLE LINEAR REGRESSION AND PETROLOGICAL MIXING MODELS

6.1 INTRODUCTION

This chapter is concerned basically with the relationship between one variable and several others and is, therefore, an extension into p-dimensional space of some of the methods of Chapter 4, such as linear regression. However, apart from principal components analysis there appears to be no general p-dimensional equivalent of structural or functional relationships described in Chapter 4. The common and related problem of multiple regression when all variables are subject to error has been discussed by Hey (1969), who showed that the relationships are best described by the eigenvectors corresponding to the smallest eigenvalues (see also section 7.6).

6.2 MULTIPLE LINEAR REGRESSION

The basic mathematical model for multiple linear regression is that:-

$$y = \alpha + \beta_1 x_1 + \beta_2 x_2 + \ldots + \beta_p x_p + \varepsilon \tag{6.1}$$

where x_1, x_2, ... x_p are the p independent variables, y is the dependent variable, α, β_1, ... β_p are constants and ε is a random variable. Obviously, when p = 1, this simplifies to the linear regression model of equation 4.13.

A point that should be realized from the start, is that the variables x_1, ... x_p do not necessarily have to be separate variables, but can be functions of the same variable. For example, to fit a cubic curve to some data, one would simply make $x_1 = w$, $x_2 = w^2$ and $x_3 = w^3$ resulting in the model:-

$$y = \alpha + \beta_1 w + \beta_2 w^2 + \beta_3 w^3 + \varepsilon \tag{6.2}$$

where w is a single independent variable. Now although this is the equation of a cubic curve, it is still a linear equation in the mathematical sense, hence this type of multiple regression is known as *curvilinear regression*. The specific example of equation 6.2 is also known as *polynomial regression*. In general, any equation that can be reorganized into a linear form can be used as the basic model. For example:-

$$y = \alpha x_1^{\beta_1} x_2^{\beta_2} \tag{6.3}$$

is not a linear equation, but by taking logarithms of both sides it can be written as:-

$$\log y = \log \alpha + \beta_1 \log x_1 + \beta_2 \log x_2 + \log \varepsilon \qquad (6.4)$$

which is a linear equation of the same type as equation 6.1. Several other equations of this type are discussed by Draper & Smith (1966).

To return to equation 6.1, for each of the n sets of observations, which should be taken from a random sample, we can write an equation of the type:-

$$y_i = \alpha + \beta_1 x_{i1} + \beta_2 x_{i2} + \ldots + \beta_p x_{ip} + \varepsilon_i \qquad (6.5)$$

where ε_i is the deviation of the ith observation from the true surface fitted by equation 6.5. As with simple linear regression, coefficients α, β_1, β_2, ... β_p are usually estimated by the method of least squares, by minimizing S, the sum of squares of the deviations, ε_i, so that the quantity to be minimized is:-

$$S = \sum \varepsilon_i{}^2 = \sum (y_i - \alpha - \beta_1 x_{i1} - \ldots - \beta_p x_{ip})^2 \qquad (6.6)$$

Differentiating this equation with respect to each coefficient individually, and equating the resulting equations to zero, provides the normal equations as:-

$$
\begin{aligned}
pa \quad &+ b_1 \sum x_{i1} \quad + b_2 \sum x_{i2} \quad + \ldots + b_p \sum x_{ip} = \sum y_i \\
a \sum x_{i1} &+ b_1 \sum x_{i1}{}^2 \quad + b_2 \sum x_{i1} x_{i2} + \ldots + b_p \sum x_{i1} x_{ip} = \sum x_{i1} y_i \\
\ldots \quad & \quad \ldots \qquad \qquad \ldots \qquad \qquad \ldots \qquad \ldots \\
a \sum x_{ip} &+ b_1 \sum x_{ip} x_{i1} + b_2 \sum x_{ip} x_{i2} + \ldots + b_p \sum x_{ip}{}^2 = \sum x_{ip} y_i
\end{aligned}
\qquad (6.7)
$$

These $p + 1$ equations must then be solved for the $p + 1$ coefficients a, b_1, ... b_p, which are the estimates of α, β_1, ... β_p. This is easily performed by a computer, but if p is large, the amount of computer time can be considerable, and on machines with small word sizes, considerable loss of precision can occur (Longely, 1967). To help minimize these problems the normal equations can be rewritten to eliminate a, the constant term, which using the notation of equations 2.3 and 4.2 produces the following set of equations:-

$$
\begin{aligned}
b_1 SSx_1 \quad &+ b_2 SPx_1 x_2 + \ldots + b_p SPx_1 x_p = SPx_1 y \\
b_1 SPx_2 x_1 &+ b_2 SSx_2 \quad + \ldots + b_p SPx_2 x_p = SPx_2 y \\
\ldots \quad & \quad \ldots \qquad \qquad \ldots \qquad \ldots \\
b_1 SPx_p x_1 &+ b_2 SPx_p x_2 \quad \ldots + b_p SSx_p = SPx_p y
\end{aligned}
\qquad (6.8)
$$

Note that when $p = 1$, these give the same result as equation 4.20 used for linear regression. These equations can also be written in full matrix form as:-

$$
\begin{bmatrix}
SSx_1 & SPx_1 x_2 & \ldots & SPx_1 x_p \\
SPx_2 x_1 & SSx_2 & \ldots & SPx_2 x_p \\
\ldots & \ldots & \ldots & \ldots \\
SPx_p x_1 & SPx_p x_2 & \ldots & SSx_p
\end{bmatrix}
\cdot
\begin{bmatrix}
b_1 \\
b_2 \\
.. \\
b_p
\end{bmatrix}
=
\begin{bmatrix}
SPx_1 y \\
SPx_2 y \\
\ldots \\
SPx_p y
\end{bmatrix}
\qquad (6.9)
$$

which is equivalent in abbreviated matrix notation to:-

$$\mathbf{X}\,\mathbf{b} = \mathbf{Y} \qquad (6.10)$$

where **X** is a symmetrical pxp matrix, known as the *sums of squares and products matrix*, and **b** and **Y** are px1 vectors of the b_i and SPx_iy_i, respectively. If both sides of the equation are pre-multiplied by $\mathbf{X}^{-1}$, the inverse of **X**, we obtain:-

$$\mathbf{X}^{-1}\mathbf{X}\,\mathbf{b} = \mathbf{X}^{-1}\mathbf{Y} \tag{6.11}$$

which, as $\mathbf{X}^{-1}\mathbf{X} = \mathbf{I}$, where **I** is a pxp identity matrix with ones in each diagonal term and zeros elsewhere, simplifies to:-

$$\mathbf{b} = \mathbf{X}^{-1}\mathbf{Y} \tag{6.12}$$

To calculate the inverse of large matrices by hand is almost impossible, but it is relatively simple if some of the readily available standard subroutines are used in a computer program. Care must be taken, however, as not all matrices have an inverse. For example, if the variables $x_1, \dots x_p$ are closed, the matrix will be *singular* and, by definition, will have no inverse. If this happens, one of the variables must be omitted to obtain a solution.

Having found the coefficients, b_i, the constant term, a, can then be found using the relationship that:-

$$a = \bar{y} - b_1\bar{x}_1 - b_2\bar{x}_2 - \dots - b_p\bar{x}_p \tag{6.13}$$

where $\bar{y}, \bar{x}_1, \dots \bar{x}_p$ are the means of the variables. For further details of the matrix approach to multiple regression the reader may refer to Koch & Link (1971), Marriott (1974), Tatsuoka (1971), or Walpole & Myers (1972).

If the independent variables $x_1, \dots x_p$ are locational values such as latitude, longitude, height or depth, the multiple regression model becomes a *trend surface*. The results are usually presented as a map in which predicted values of the dependent variable are plotted as contours, which may be planar surfaces, polynomial surfaces or in more sophisticated examples, surfaces representing double Fourier series. In some trend surface work more emphasis is placed upon the deviations from the fitted surface than on the surface itself. For example, with geochemical data, where one is looking for anomalies, the surface can be regarded as representing the regional trend and large deviations from the surface as the anomalies. Much has been written on these topics and an excellent summary of the methods available, together with some of their problems, is given by Davis (1973).

Jones (1972) points out that care should be taken if the independent variables are correlated, which is often the case with polynomial trend surfaces, as such data can give rise to highly unstable regression coefficients. Under such circumstances he advocates the use of a technique called *ridge-regression*.

Numerical example

As a simple example of multiple linear regression, we shall again use the cationic K% data of Hutchison (1975) for the 17 volcanoes on Java, which we will assume to be a random sample. However, instead of using the depth to the

Benioff zone, we shall use the latitudes and longitudes of the volcanoes as the two independent variables, and cationic K% as the dependent variable. The data and calculations are set out in Table 6.1. As can be seen the K% can be estimated by the equation:-

K% = 0.4518(Longitude - 100) - 1.9154(Latitude) + 11.1535

This planar surface, whose strike is remarkably similar to that of the supposed Benioff zone, is shown in Fig. 6.1.

TABLE 6.1

Worked example of multiple linear regression using the latitude and longitude of 17 volcanoes form Java as two independent variables, and the cationic K% at Si% = 51 as the dependent variable, taken from the data of Hutchison (1975) which are assumed to be a random sample. For the location of the volcanoes see Fig. 6.1. To reduce the magnitude of some of the numbers, 100 has been subtracted from the longitude values.

Name of volcano	K% (y)	Longitude (x_1)	Latitude (x_2)
1. Krakatau	1.62	5.42	6.10
2. Tangkuban	0.98	7.60	6.77
3. Papandajan	0.24	7.73	7.32
4. Guntur	1.08	7.33	7.15
5. Galunggung	0.71	8.05	7.25
6. Tjerimai	1.37	8.40	6.90
7. Slamet	1.74	9.22	7.25
8. Dieng	2.39	9.92	7.20
9. Sumbing	0.86	10.07	7.38
10. Ungaran	3.27	10.33	7.18
11. Merbabu	1.95	10.43	7.45
12. Merapi	2.29	10.45	7.55
13. Kelud	1.23	12.32	7.93
14. Semeru	1.38	13.58	7.95
15. Bromo	2.15	12.95	7.95
16. Raung	2.18	14.05	8.13
17. Batur	1.65	15.38	8.25

$\sum y_i = 27.09$ $\sum x_{1i} = 173.23$ $\sum x_{2i} = 125.71$

$\sum y_i^2 = 51.9233$ $\sum x_{1i}^2 = 1883.3449$ $\sum x_{2i}^2 = 934.2939$

$\sum x_{1i}x_{2i} = 1302.9082$ $\sum x_{1i}y_i = 287.4259$ $\sum x_{2i}y_i = 201.2146$

$\bar{y} = 1.5935$ $\bar{x}_1 = 10.1900$ $\bar{x}_2 = 7.3947$

Equations 2.3 and 4.2 then give the following matrices:-

$$\mathbf{X} = \begin{bmatrix} 118.131200 & 21.923300 \\ 21.923300 & 4.705424 \end{bmatrix} \qquad \mathbf{Y} = \begin{bmatrix} 11.37880 \\ 0.892018 \end{bmatrix}$$

from which $\mathbf{X}^{-1}$ and **b** (using equation 6.12) may be obtained:-

$$\mathbf{X}^{-1} = \begin{bmatrix} 0.062550 & -0.291432 \\ -0.291432 & 1.570346 \end{bmatrix} \quad \text{and} \quad \mathbf{b} = \begin{bmatrix} 0.4518 \\ -1.9154 \end{bmatrix}$$

Equation 6.13 then gives the constant term as:-

a = 1.5935 - (0.4518 x 10.19) + (1.9154 x 7.3947) = 11.1535

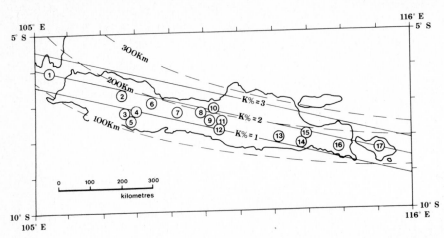

Fig. 6.1. Map of Java, based mainly on Hutchison (1975), with dashed curves showing the supposed location of the Benioff zone. The solid lines are the contours of cationic K% for values of 1, 2 and 3, as defined by the multiple regression equation K% = 0.4518(Long – 100) – 1.9154(Lat) + 11.1535 from Table 6.1. The key to the circled numbers locating each volcano given in Table 6.1 is:- 1 = Krakatau; 2 = Tangkuban; 3 = Papandajan; 4 = Guntur; 5 = Galunggung; 6 = Tjerimai; 7 = Slamet; 8 = Dieng; 9 = Sumbing; 10 = Ungaran; 11 = Merbabu; 12 = Merapi; 13 = Kelud; 14 = Semeru; 15 = Bromo; 16 = Raung; 17 = Batur.

6.2.1 *Testing the significance of the regression*

As with linear regression, the overall test of significance of the regression can be performed with an analysis of variance. The total sum of squares is again partitioned into two parts, one attributable to the regression and the other to the random errors. Using matrix notation the sum of squares due to the regression, S_R, which has p d.f., is given by:-

$$S_R = b^* Y \qquad (6.14)$$

where b^* is the transpose of b. As with linear regression, the total sum of squares has n – 1 d.f. and is given by $S_T = SS_y$. The generalized analysis of variance is set out in Table 6.2.

Another commonly quoted measure of the adequacy of a fitted regression is called the *coefficient of multiple determination*, R^2, and is given by:-

$$R^2 = S_R / S_T \qquad (6.15)$$

i.e. it is simply the proportion of the sum of squares accounted for by the regression. R is called the *multiple correlation coefficient*.

One of the problems of multiple regression is knowing how many variables to include – this is particularly true with high order polynomials, as discussed by Chayes (1970d) with respect to trend surfaces. Fortunately, an analysis of variance can be performed to test the significance of including extra variables

TABLE 6.2

Generalized computational scheme for the analysis of variance to test the significance of multiple linear regression, together with a numerical example using the data of Table 6.1

ANALYSIS OF VARIANCE: MULTIPLE LINEAR REGRESSION

Source of variation	Sum of Squares	Degrees of freedom	Variance or mean square	F-ratio
Regression	$S_R = \mathbf{b^*Y}$	p	$s_r^2 = S_R/p$	$F = s_r^2/s_e^2$
Error	$S_E = S_T - S_R$	n-p-1	$s_e^2 = S_E/(n-p-1)$	
Total	$S_T = SS_y$	n-1		

NUMERICAL EXAMPLE

From Table 6.1
$S_R = (11.3788 \times 0.4518) - (0.892018 \times 1.9154) = 3.4324$
$S_T = 51.9233 - 27.092/17 = 8.7546$

Source of variation	Sum of Squares	Degrees of freedom	Variance or mean square	F-ratio
Regression	3.4324	2	1.7162	4.51
Error	5.3222	14	0.3802	
Total	8.7546	16		

Conclusion: F = 4.51 is significant as $F_{0.05} = 3.74$ with 2 and 14 d.f.

in the regression. Suppose that a new multiple regression is performed with some extra variables, so that the total number is now q instead of p. The analysis of variance as shown in Table 6.2 will only test the overall significance of the regression, so a different approach has to be taken. If S_R is the sum of squares due to the regression of the p variables, and S_R' is the sum of squares due to the regression of the q variables, the sum of squares due to the additional variables is given by $S_R' - S_R$, and has q - p d.f. The F-ratio to be tested, F_a, is the mean square, s_a^2, divided by, s_e^2, the error mean square of the regression using the q variables, as set out in Table 6.3. If q = p - 1 this gives the same result as the test of significance for b_q, the qth regression coefficient, as described in the next section. However, to test all possible combinations of variables is tedious and several methods, which are critically discussed by Draper & Smith (1966), have been suggested for avoiding this e.g. backward elimination and foward selection procedures, stepwise and stagewise regression.

Marriott (1974) also deals with this problem and besides discussing some of the more theoretical aspects of deciding which variables should be included in a multiple regression, offers some practical advice. For example, he suggests

TABLE 6.3

Generalized computational scheme for an analysis of variance to test the significance of using additional variables in multiple linear regression, together with a numerical example.

ANALYSIS OF VARIANCE: MULTIPLE LINEAR REGRESSION

(ADDITIONAL VARIABLES)

Source of variation	Sum of Squares	Degrees of freedom	Variance or mean square	F-ratio
Regression 2	$S_R' = \mathbf{b}_2 * \mathbf{Y}_2$	q	$s_r^2 = S_R'/q$	$F = s_r^2/s_e^2$
Regression 1	$S_R = \mathbf{b}_1 * \mathbf{Y}_1$	p		
Extra variables	$S_A = S_R' - S_R$	q-p	$s_{a2} = S_A/(q-p)$	$F_a = s_{a2}/s_{e2}$
Error	$S_E = S_T - S_R'$	n-q-1	$s_{e2} = S_E/(n-q-1)$	
Total	$S_T = SS_y$	n-1		

NUMERICAL EXAMPLE

If the data of Table 6.1 are used to fit a 2nd order polynomial to the K% data, with x_1, x_2, x_1x_2, x_1^2 and x_2^2 as independent variables (q = 5), the sum of squares due to the regression is 5.0837.

Source of variation	Sum of Squares	Degrees of freedom	Variance or mean square	F-ratio
Regression 2	5.0837	5	1.0167	3.05
Regression 1	3.4324	2		
Extra variables	1.6513	3	0.5504	1.65
Error	3.6709	11	0.3337	
Total	8.7546	16		

Conclusions: Additional variables - $F_a = 1.65$ is not significant as $F_{0.05} = 3.59$ with 3 and 11 d.f.   Overall regression - $F = 3.05$ is not significant as $F_{0.05} = 3.20$ with 5 and 11 d.f.

that it is better to include irrelevant variables rather than exclude important ones, and if two variables are highly correlated, the easier or cheaper one to measure would be the more logical choice.

Numerical example
An example of the analysis of variance of a multiple linear regresssion is given in Table 6.2, using the data of Table 6.1 which are assumed to be a random sample. At the 95% confidence level, the regression is judged to be significant, as Table A4 in the Appendix gives $F_{0.05} = 3.74$ with 2 and 14 d.f., which is less than the calculated value of 4.51. The coefficient of multiple determination, R^2, is $3.4324/8.7546 = 0.392$ and the multiple correlation coefficient, R, is therefore 0.626.
Table 6.3 also gives a worked example of the analysis of variance to test the significance of introducing 3 additional variables, x_1x_2, x_1^2 and x_2^2, to the

regression to make it a 2nd degree polynomial. As can be seen, with an F-ratio of only 1.65, there is no evidence to suggest that the 3 extra variables contribute significantly to the regression, as Table A4 in the Appendix gives $F_{0.05} = 3.59$ with 3 and 11 d.f. Furthermore, there is no evidence to suggest that the 2nd degree regression as a whole is significant at the 95% confidence level, as the calculated F-ratio of 3.05, is less than the critical value of $F_{0.05} = 3.20$ with 5 and 11 d.f.

6.2.2 Confidence limits for the regression coefficients

Confidence limits for the regression coefficients are easily obtained as x^{ij}, the individual terms of the inverse matrix $\mathbf{X}^{-1}$, are proportional to the variances and covariances of the coefficients. The exact relationship is given by:-

$$\text{var } b_i = x^{ii} s_e^2 \tag{6.16}$$

where s_e^2 is the mean error square as defined in Table 6.2. Similarly, the covariance between the ith and jth coefficients is given by:-

$$\text{cov } b_i b_j = x^{ij} s_e^2 \tag{6.17}$$

The $100(1 - \alpha)\%$ confidence limits for any coefficient can then be expressed as:-

$$b_i \pm t_{\alpha/2} \sqrt{(x^{ii} s_e^2)} \tag{6.18}$$

where $t_{\alpha/2}$ is the value of Student's t with $n - p - 1$ d.f. obtained from Table A3 in the Appendix.

Numerical example

Using the data of Tables 6.1 and 6.2 and equation 6.18, the 95% confidence limits for the coefficient b_1, are given by:-

$$0.4518 \pm 2.145 \sqrt{(0.06225 \times 0.3802)} \quad \text{or} \quad 0.45 \pm 0.33$$

assuming that the data are a random sample. Similarly, the 95% confidence limits for b_2 are -1.9 ± 1.7. Note that at the 95% confidence level, both of these coefficients are significant i.e. they are not zero, which confirms the overall test of significance made in the last example. However, in the case of the 2nd order polynomial regression, the coefficients and their 95% confidence limits are:-

$b_1 = 14 \pm 25$ $b_2 = -70 \pm 130$ $b_3 = -2.3 \pm 4.4$
$b_4 = 0.15 \pm 0.40$ $b_5 = 6 \pm 12$

which indicates that all of them could be zero i.e. the overall 2nd order polynomial regression is not significant.

6.2.3 Confidence limits for the location of the regression surface

The variance of $\hat{y}$, when $\hat{y}$ is an estimate of the value of the regression "surface", is given by:-

$$s_y^2 = s_e^2 \left[1/n + (\mathbf{x} - \bar{\mathbf{x}})^* \mathbf{X}^{-1}(\mathbf{x} - \bar{\mathbf{x}}) \right] \qquad (6.19)$$

where n is the number of observations, s_e^2 is the error mean square as defined in Table 6.2, $\mathbf{x}$ is the vector of values at which y is to be estimated, $\bar{\mathbf{x}}$ is the vector of mean values of the independent variables, * stands for transpose and $\mathbf{X}^{-1}$ is the inverse matrix used in equation 6.11. Note that if the number of independent variables is one, equation 6.19 reduces to equation 4.31 used for linear regression. The $100(1 - \alpha)\%$ confidence limits can then be written as:-

$$\hat{y} \pm t_{\alpha/2} \, s_y \qquad (6.20)$$

where $t_{\alpha/2}$ is the value of Student's t with $n - p - 1$ d.f. obtained from Table A3 in the Appendix.

Numerical example

Using the data of Table 6.1, we will first determine the 95% confidence limits on y, the cationic K%, at $x_1 = 10.19$ and $x_2 = 7.3947$ i.e. the mean values of the independent variables, where $y = \bar{y} = 1.5935$. Now as $[\mathbf{x} - \bar{\mathbf{x}}] = [0.0 \quad 0.0]$ it follows that $(\mathbf{x} - \bar{\mathbf{x}})^* \mathbf{X}^{-1}(\mathbf{x} - \bar{\mathbf{x}}) = 0.0$. Using the value of s_e^2, from Table 6.2, and equation 6.19 we then obtain:-

$$s_y^2 = 0.3802 \, (1/17 + 0.0) = 0.022365 \quad \text{and} \quad s_y = 0.149548$$

Table A3 in the Appendix gives $t_{0.025} = 2.145$ with 14 d.f., so that the 95% confidence limits on y are $1.59 \pm 2.145 \times 0.149548$ or 1.6 ± 0.3 assuming, of course, that the data of Table 6.1 are a random sample.

We will now estimate the 95% confidence limits on y at $x_1 = 7.19$ and $x_2 = 6.3947$. Using the regression coefficients determined in Table 6.1, $\hat{y}$ is given by:-

$$\hat{y} = 11.1535 + (0.4518 \times 7.19) - (1.9154 \times 6.3947) = 2.15$$

We next evaluate the vector $(\mathbf{x} - \bar{\mathbf{x}})$ as:-

$$\begin{bmatrix} 7.19 - 10.19 & 6.3947 - 7.3947 \end{bmatrix} = \begin{bmatrix} -3 & -1 \end{bmatrix}$$

which gives $(\mathbf{x} - \bar{\mathbf{x}})^* \mathbf{X}^{-1}(\mathbf{x} - \bar{\mathbf{x}})$ as:-

$$\begin{bmatrix} -3 & -1 \end{bmatrix} \cdot \begin{bmatrix} 0.062550 & -0.291432 \\ -0.291432 & 1.570346 \end{bmatrix} \cdot \begin{bmatrix} -3 \\ -1 \end{bmatrix}$$

$$= \begin{bmatrix} 0.103782 & -0.696050 \end{bmatrix} \cdot \begin{bmatrix} -3 \\ -1 \end{bmatrix} = 0.384704$$

Equation 6.19 then gives s_y^2 as:-

$s_y^2 = 0.3802 \ (1/17 + 0.384704) = 0.168629$ and $s_y = 0.410645$

which gives the 95% confidence limits as $2.15 \pm 2.145 \times 0.410645$ or 2.2 ± 0.9.

6.2.4 *Confidence limits for an individual value of y*

By analogy with linear regression, the variance of $\hat{y}$, when $\hat{y}$ is an estimate of an individual value of y, is given by the expression:-

$$s_y^2 \ = \ s_e^2 \left[1 + 1/n + (\mathbf{x} - \bar{\mathbf{x}})^* \mathbf{X}^{-1} (\mathbf{x} - \bar{\mathbf{x}}) \right] \tag{6.21}$$

and confidence limits can then be obtained by using equation 6.20.

Numerical example

Using the data of the previous example, at $x_1 = 10.19$ and $x_2 = 7.3947$ we obtain:-

$s_y^2 = 0.3802 \ (1 + 1/17 + 0.0) = 0.402565$ and $s_y = 0.634480$

so that the 95% confidence limits for an individual value of y are given by $1.59 \pm 2.145 \times 0.634480$ or 1.6 ± 1.4. Similarly, at $x_1 = 7.19$ and $x_2 = 6.3947$ we have:-

$s_y^2 = 0.3802 \ (1 + 1/17 + 0.384704) = 0.548829$ and $s_y = 0.740830$

which gives the confidence limits as $2.15 \pm 2.145 \times 0.74083$ or 2.2 ± 1.6.

6.3 PETROLOGICAL MIXING MODELS

In petrology it is often useful to be able to estimate the proportions of one set of chemical compositions, such as mineral analyses, that best approximate another set of compositions, such as one rock analysis in the simplest case. Such problems have become known as petrological mixing models and over the past decade many methods have been proposed for solving them (Albarède & Provost, 1977; Banks, 1979; Bryan, 1969; Bryan et al., 1969a, 1969b; Chayes, 1968, 1970c; Gray, 1973; Greenwood, 1967, 1968; Le Maitre, 1979; Perry, 1967; Provost & Allègre, 1979; Reid et al., 1973; Stormer & Nicholls, 1978; Wright & Doherty, 1970). Of all the techniques described in this book, it is probably fair to state that petrological mixing models have recieved wider acceptance and use in petrology than any others. Basically the petrological mixing models are similar to multiple regression, but with two important differences. Firstly, no constant term is involved, and secondly, what is usually regarded as a variable in multiple regression is an observation in petrological mixing models and *vice versa*. In the jargon of factor analysis (Chapter 8), classic multiple regression is R-mode and petrological mixing models are Q-mode, and although this makes no difference to the estimation of the proportions, which are the regression coefficients, it does affect tests of significance.

6.3.1 *The simple mixing model*

The simple mixing model involves the following type of problem. Given the chemical compositions of a bulk rock and all its mineral phases, how do we estimate the mode of the rock? If there are p oxides (variables) and n mineral phases (observations), we can then write for each of the p oxides that:-

$$y_1 = b_1x_{11} + b_2x_{21} + \text{.....} + b_nx_{n1}$$
$$y_2 = b_1x_{12} + b_2x_{22} + \text{.....} + b_nx_{n2}$$
$$\text{...} \qquad \text{...} \qquad \text{...} \qquad \text{...} \qquad \text{...}$$
$$y_p = b_1x_{1p} + b_2x_{2p} + \text{.....} + b_nx_{np}$$

$$(6.22)$$

where y_i is the ith oxide of the bulk composition, $x_{1i}, \text{...} x_{ni}$ are the ith oxides of the n mineral phases, and $b_1, \text{...} b_n$ are the coefficients (mode) to be estimated. This set of equations can also be written in matrix form as:-

$$(6.23)$$

$$\mathbf{y} = b_1\mathbf{x}_1 + b_2\mathbf{x}_2 + \text{....} + b_n\mathbf{x}_n$$

where $\mathbf{y}, \mathbf{x}_1, \mathbf{x}_2, \text{..} \mathbf{x}_n$ are the composition vectors.

The simple model can also be used for problems involving differentiation e.g. "ankaramite = basalt + certain mineral phases" or "andesite = basalt – certain mineral phases". In the latter case, the coefficients for the mineral phases would be expected to be negative in any solution. As will be seen in section 6.3.3, the solution to the problem "basalt = andesite + certain mineral phases" will, in general, give different proportions to the solution "andesite = basalt – certain mineral phases", so that it is important to express the problem in the correct form, with the composition to be estimated as the y_i's.

Now unless n, the number of unknowns, is equal to p, the number of oxides, the equations 6.22 cannot be solved as a set of simultaneous linear equations. Perry (1967) overcame this problem by replacing certain mineral phases by their end-member compositions e.g. Ab and An instead of plagioclase, until n = p, when the equations can be solved directly. However, as Wright & Doherty (1970) point out, unstable proportions may be obtained unless the amount of the split phase is likely to be more than about 10%. From personal experience this may also happen if very similar compositions are included (e.g. An_{55} and An_{65} for the rim and core of a plagioclase) when a large positive proportion for one may be accompanied by a large negative proportion for the other.

However, as Bryan (1969), Bryan et al. (1969a, 1969b) and Chayes (1968) show, as p is invariably greater than n (or should be according to the Phase rule), the system is over-determined so that the coefficients may be solved by the method of least squares, by minimizing S where:-

$$S = \sum (y_i - \hat{y}_i)^2 = \sum (y_i - b_1x_{1i} - b_2x_{2i} - \text{...} - b_nx_{ni})^2 \qquad (6.24)$$

Differentiating this with respect to each coefficient in turn and equating the results to zero, then yields the normal equations as:-

$$b_1 \sum x_{1i}^2 + b_2 \sum x_{1i}x_{2i} + \ldots\ldots + b_n \sum x_{1i}x_{ni} = \sum x_{1i}y_i$$
$$b_1 \sum x_{2i}x_{1i} + b_2 \sum x_{2i}^2 + \ldots\ldots + b_n \sum x_{2i}x_{ni} = \sum x_{2i}y_i$$
$$\ldots\ldots \qquad \ldots\ldots \qquad \ldots\ldots \qquad \ldots\ldots$$
$$b_1 \sum x_{ni}x_{1i} + b_2 \sum x_{ni}x_{2i} + \ldots\ldots + b_n \sum x_{ni}^2 = \sum x_{ni}y_i \qquad (6.25)$$

Note that these equations are similar to equations 6.7, except that the first row and column are missing, due to the absence of the constant term in equations 6.22, and that the summations are made over the oxides rather than the observations.

As these equations are identical in form to equations 6.8, they can then be solved for the b_i's using the matrix approach of equation 6.12.

Numerical example

To illustrate the simple mixing model, we will use three hypothetical feldspar compositions. It is proposed to see what proportions of the two feldspars $F_1 = Or_5Ab_5An_{90}$ and $F_2 = Or_{80}Ab_{15}An_5$, best approximate the feldspar $F = Or_{10}Ab_{45}An_{45}$ i.e. to test the model $F = b_1F_1 + b_2F_2$. Using equation 6.25 to form the terms required for the normal equations we have:-

$$\sum x_{1i}^2 = 5^2 + 5^2 + 90^2 = 8150$$
$$\sum x_1x_{2i} = (5 \times 80) + (5 \times 15) + (90 \times 5) = 925$$
$$\sum x_{2i}^2 = 80^2 + 15^2 + 5^2 = 6650$$
$$\sum x_{1i}y_i = (10 \times 5) + (45 \times 5) + (45 \times 90) = 4325$$
$$\sum x_{2i}y_i = (10 \times 80) + (45 \times 15) + (45 \times 5) = 1700$$

which gives the normal equations in matrix form as:-

$$\begin{bmatrix} 8150 & 925 \\ 925 & 6650 \end{bmatrix} \cdot \begin{bmatrix} b_1 \\ b_2 \end{bmatrix} = \begin{bmatrix} 4325 \\ 1700 \end{bmatrix}$$

Taking the inverse and rewriting this in the form of equation 6.12 gives:-

$$\mathbf{b} = \begin{bmatrix} 1.2467 \times 10^{-4} & -1.7341 \times 10^{-5} \\ -1.7341 \times 10^{-5} & 1.5279 \times 10^{-4} \end{bmatrix} \cdot \begin{bmatrix} 4325 \\ 1700 \end{bmatrix}$$

from which we can obtain:-

$$b_1 = (1.2467 \times 10^{-4} \times 4325) + (-1.7341 \times 10^{-5} \times 1700) = 0.5097 \quad \text{and}$$
$$b_2 = (-1.7341 \times 10^{-5} \times 4325) + (1.5279 \times 10^{-4} \times 1700) = 0.1847$$

The estimated best-fit "composition", $\hat{F}_u$, is then found using equation 6.22 and gives:-

$$\hat{y}_1 \text{ (Or)} = (0.5097 \times 5) + (0.1847 \times 80) = 17.32$$
$$\hat{y}_2 \text{ (Ab)} = (0.5097 \times 5) + (0.1847 \times 15) = 5.32$$
$$\hat{y}_3 \text{ (An)} = (0.5097 \times 90) + (0.1847 \times 5) = 46.80$$

Note that as the coefficients only sum to 0.6944, the feldspar components sum to 69.44, which means that the estimated best-fit "composition" is not, strictly speaking, a valid composition. The residual sum of squares can then be calculated as:-

$$S = \sum (y_i - \hat{y}_i)^2 = (10 - 17.32)^2 + (45 - 5.32)^2 + (45 - 46.80)^2 = 1631.32$$

and the square root of this value, 40.39, is the distance in the 3-dimensional feldspar space between the estimated "composition", F_u, and the actual composition, F, which indicates that the fit is far from good! Fortunately, this is only an example chosen to illustrate the geometry of the fit, as will be seen in section 6.3.3.

If the estimated "composition" $\hat{F}_u$ is recalculated to 100% to give F_r, as is often done, we obtain a composition Or = 24.94, Ab = 7.66 and An = 67.40, but the residual sum of squares then becomes 2119.24, which is much greater than the previous value.

6.3.2 The constrained simple mixing model

One limitation of the method just described is that the coefficients, b_i, can assume any positive or negative values. Now in some types of petrological mixing problems, both positive and negative values of b_i are physically meaningful, e.g. in certain metasomatic and differentiation processes, where some materials may be added, while others are being removed. In other mixing problems, however, only positive or zero values have any physical meaning e.g. in determining the mode of a rock.

Wright & Doherty (1970) and Banks (1979) argue that as negative values of b_i are physically meaningless in certain mixing models (e.g. in estimating rock modes), linear programming methods should be used as these have provisions for constraining the coefficients to be positive (see section 6.3.5). However, the constraint of only allowing non-negative proportions does not have universal acceptance. As Stormer & Nicholls (1978) point out, and the author agrees, there is much to be said for *not* constraining the b_i in this way "as unexpected negative values may provide insight into the actual process". This is particularly true for computer programs which can be run interactively such as XLFRAC (Stormer & Nicholls, 1978), LESQ02 in CLAIR (Le Maitre & Ferguson, 1978) and GENMIX (Le Maitre, 1981), when the user rapidly acquires a feel for the data, and the effect that particular compositions have upon the fit. Reid et al. (1973) also discuss the application of this constraint and comment that "large negative or other non-physically meaningful values ... ... implies that an error(s) has been made in the definition of the problem" and conclude that the "application of 'non-negativity' constraints without examination of the unconstrained estimates may be dangerous".

However, for data which have a constant sum, such as major element data, there is a far more important constraint which is inherent in the nature of the problem, which is that the coefficients should sum to 1 or:-

$$\sum b_i = 1 \tag{6.26}$$

The reason for the importance of this constraint is simple. As mixing problems of the type defined by equations 6.22 are basically predicting compositions, all the predictions must lie in the $(p - 1)$ hyperplane of constant composition (see Fig. 5.1 or 6.2). However, it follows from equations 6.22 that if all the analyses sum to 100%, the predicted composition will only sum to 100% if the sum of the b_i's is 1. If the predicted composition does not sum to 100% it is not strictly speaking a valid composition. Admittedly the $\hat{y}_i$'s can always be, and often are, recalculated to sum to 100%, but this only increases the residual sum of squares $\sum (y_i - \hat{y}_i)^2$ and "worsens" the fit - a point also noted by Reid et al. (1973).

Equation 6.22 can be rewritten to include the constraint that $\sum b_i = 1$ as follows:-

$$\mathbf{y} = b_1\mathbf{x}_1 + b_2\mathbf{x}_2 + \ldots + b_r\mathbf{x}_r + (1 - b_1 - b_2 - \ldots - b_r)\mathbf{x}_n \tag{6.27}$$

where $r = p - 1$. The quantity to be minimized for the least squares solution is then:-

$$S = \sum \left[y_i - b_1x_{1i} - \ldots - b_rx_{ri} - (1 - b_1 - b_2 - \ldots - b_r)x_{ni} \right]^2$$
$$= \sum (u_i - b_1z_{1i} - \ldots - b_rz_{ri})^2 \tag{6.28}$$

where $u_i = y_i - x_{ni}$, and $z_{ji} = x_{ji} - x_{ni}$. Equation 6.28 is now identical in form to equation 6.24, but has one less term, and can be solved for the b_i in exactly the same manner, using the normal equations of 6.25 with the x's and y's replaced with the z's and u's, respectively. Although the r values of b_i will be unconstrained, the constraint is reimposed when b_n is calculated as:-

$$b_n = 1 - \sum_{i=1}^{r} b_i \tag{6.29}$$

Williams (1959) gives an alternative method of dealing with constrained coefficients, in which they are derived from the unconstrained coefficients. This has the advantage that it is easy to see the effect of the constraint.

Numerical examples

Case 1. Using the three feldspar compositions of the last example we will now calculate the solution to the simple mixing model $F = b_1F_1 + b_2F_2$, subject to the constraint that $b_1 + b_2 = 1$. With only two coefficients the solution is trivial as the normal equations reduce to a single equation:-

$$b_1 \sum z_{1i}^2 = \sum z_{1i}u_i$$

Substituting values we obtain:-

$$\sum z_{1i}^2 = (5 - 80)^2 + (5 - 15)^2 + (90 - 5)^2 = 12950$$
$$\sum z_{1i}u_i = (5 - 80)(10 - 80) + (5 - 15)(45 - 15) + (90 - 5)(45 - 5) = 8350$$

which then gives $b_1 = 8350/12950 = 0.6448$ and $b_2 = 1 - 0.6448 = 0.3552$. The estimated composition $\hat{F}_c$ is then $\hat{y}_1(Or) = (0.6448 \times 5) + (0.3552 \times 80) = 31.64$ and similarly $\hat{y}_2(Ab) = 8.55$ and $\hat{y}_3(An) = 59.81$. The residual sum of squares is:-

$$(10 - 31.64)^2 + (45 - 8.55)^2 + (45 - 59.81)^2 = 2016.23$$

which is between the values of the residual sum of squares for the two solutions of the example in the previous section.

   Case 2.   On a more realistic level, let us look at a numerical example of one of the more important uses of the constrained simple mixing model, i.e. computing the modes of rocks. Table 6.4 gives some unpublished data, obtained

TABLE 6.4

Example of the use of the constrained simple mixing model to estimate the mode of the type dacite from Poiene, Vlădeasa Mts., Transylvania, Roumania. The bulk rock was analysed by x-ray fluorescense and all the minerals, except quartz, were analysed by electron microprobe. Abbreviations used are:- Plag. = plagioclase;   Horn. = hornblende;   Biot. =   biotite;   Alk.Feld. = alkali feldpsar; Timag. = titanomagnetite; Qtz. = quartz. FeO* is total iron as FeO.

| | Estimated | | Phenocryst | | | Groundmass | | |
	Dacite $(\hat{y})$	Dacite (y)	Plag. (x_1)	Horn. (x_2)	Biot. (x_3)	Alk.Feld (x_4)	Timag. (x_5)	Qtz. (x_6)
SiO_2	67.43	67.39	59.92	47.49	36.58	64.18	0.18	100.0
TiO_2	0.48	0.71	0.02	1.43	5.04	0.01	11.45	0.0
Al_2O_3	15.25	15.16	24.36	6.55	13.34	18.63	0.88	0.0
FeO*	3.42	3.34	0.23	13.78	16.18	0.14	77.99	0.0
MnO	0.07	0.05	0.02	0.37	0.13	0.02	0.69	0.0
MgO	1.85	1.76	0.00	14.47	13.88	0.00	0.47	0.0
CaO	3.19	3.24	6.62	11.05	0.00	0.11	0.03	0.0
Na_2O	3.88	3.92	7.36	1.51	0.55	3.75	0.02	0.0
K_2O	3.03	3.04	0.73	0.48	9.14	11.01	0.00	0.0
Total	98.60	98.61	99.26	97.13	94.84	97.85	91.71	100.0

Results using the analyses above
Estimated % mode   43.26   3.25   9.25   16.95   1.63   25.66
Residual sum of squares, $S^2 = 0.0815$
The value of 100a in equation 6.45, for which $S = 1.96 SD_S$ is 0.58

Results with the above analyses *recalculated to 100%*
Estimated % mode   43.54   3.16   8.94   16.80   1.51   26.05
Residual sum of squares, $S^2 = 0.0850$
The value of 100a in equation 6.45, for which $S = 1.96 SD_S$ is 0.54

at the University of Melbourne, that were used to calculate the mode of a specimen of the type dacite from Poiene, Vlădeasa Mts., Transylvania, Roumania, kindly supplied by Prof. A. Streckeisen. The bulk rock was analysed by x-ray fluorescence analysis and all the actual minerals, except the quartz, were analysed by electron microprobe. The mode obtained using the program GENMIX (Le Maitre, 1981) is also given in Table 6.4. As can be seen, the fit is extremely good as the sum of squares of the residuals is only 0.0815. In spite of the fact that several of the analyses sum to considerably less than 100%, the effect of recalculating the analyses to 100% before use is small. The residual sum of squares only changes to 0.0850 and the maximum change in the modal percentages is only 0.39. In many cases, the simple mixing model approach is the only reliable way of obtaining the mode of volcanic rocks, and hence could be extremely useful in implementing the recommendations of the IUGS Subcommission on the Systematics of Igneous rocks (Streckeisen, 1978) for the classification of igneous rocks.

6.3.3 *The geometric interpretation of simple models*

The subtle differences between the two simple mixing models can easily be seen geometrically (Le Maitre, 1979). Inspection of equation 6.23, used for the unconstrained simple model, shows it to be the equation of a hyperplane passing through the n mineral phases and the origin, as when $b_i = 1$ and all the other b's are zero, $\mathbf{y} = \mathbf{x}_i$, and when all the b's are zero, $\mathbf{y} = \mathbf{0}$. Similarly, as was shown in section 5.6, equation 6.27, used for the constrained simple model, is the equation of a hyperplane passing through the n mineral phases only.

Now by minimizing the sum of squares of the differences between the actual composition $\mathbf{y}$, and the estimated composition $\hat{\mathbf{y}}$, the least squares method is calculating the shortest or perpendicular distance between $\mathbf{y}$ and the hyperplane. This is illustrated in 3-dimensions in Fig. 6.2 using the feldspar compositions of the two previous numerical examples. As the Or, Ab and An components of the three feldspars all sum to 100%, F, F_1 and F_2 all lie in the Or-Ab-An plane of constant composition. The least squares estimate of the simple unconstrained model $F = b_1 F_1 + b_2 F_2$ gives $\hat{F}_u$, where $F\hat{F}_u$ is perpendicular to the plane $F_1 F_2 O$. As $\hat{F}_u$ lies between the Or-Ab-An plane and the origin O, the sum of the components will be less than 100%. As also noted by Reid et al. (1973), this is an important point, as with unconstrained models the common practice of normalizing the analyses to sum to 100% will not automatically ensure that the estimated proportions will sum to 1. $\hat{F}_u$ recalculated to 100% gives the composition F_r, which lies on the line $F_1 F_2$. The constrained solution to the model gives $\hat{F}_c$, where $F\hat{F}_c$ is perpendicular to the line $F_1 F_2$. It should now be clear that, in general, F_r will not be the same as $\hat{F}_c$ and that for the three distances:-

$$F\hat{F}_u \leqslant F\hat{F}_c \leqslant FF_r$$

(6.30)

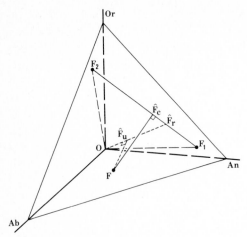

Fig. 6.2. Diagram to illustrate the geometry of the differences between the unconstrained and constrained simple mixing models. F, F_1 and F_2 represent the three feldspar compositions lying in the Or-Ab-An plane as used in the numerical example. The unconstrained solution to the model $F = b_1F_1 + b_2F_2$ gives $\hat{F}_u$ as the best estimate of F, where $F\hat{F}_u$ is perpendicular to the plane F_1F_2O. Recalculated to 100% the "composition" $\hat{F}_u$ gives F_r. The constrained solution to the same model gives $\hat{F}_c$ as the best estimate of F, where $F\hat{F}_c$ is perpendicular to the line F_1F_2.

Now as these distances are a measure of the "goodness-of-fit", being the square root of the residual sum of squares, it follows that although the unconstrained solution is generally "better" than the constrained solution, the constrained solution is generally "better" than the unconstrained solution recalculated to 100%. It should also be apparent from Fig. 6.2 that as F approaches the line F_1F_2, the difference between the two methods become smaller, until when F lies on F_1F_2, there is no difference between the two methods and the residual sum of squares becomes zero.

Summarizing the conclusions in p-dimensional space it follows that:-

the unconstrained least squares solution to the simple model gives the perpendicular from **y** *onto the n-flat defined by the n compositions and the origin.* (6.31)

and

the constrained least squares solution to the simple model gives the perpendicular from **y** *onto the (n - 1)-flat defined by the n compositions only.* (6.32)

It is important that the petrologist decides which of these two solutions is physically meaningful for the problem being solved.

It should now also be obvious that the answer obtained depends upon how the model is written. For example, it can be seen in Fig. 6.3, that with the constrained simple model, the minimized sum of squares obtained for the two

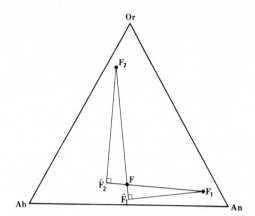

Fig. 6.3. Diagram to illustrate the geometry of the two alternative models using the same feldspars as in Fig. 6.2. $\hat{F}_2$ is the best estimate of F_2 for the constrained model $F_2 = b_1F + b_2F_1$, and $\hat{F}_1$ is the best estimate of F_1 for the model $F_1 = b_1F + b_2\hat{F}_2$. Note that the "goodness-of-fit" of the models as measured by the residual sum of squares, or the square of the distances $F_2\hat{F}_2$ and $F_1\hat{F}_1$, are not the same.

alternative models $F_1 = b_1F + b_2F_2$ and $F_2 = b_1F + b_2F_1$ will be completely different. Furthermore, the reader may care to check, either by calculation or by using Fig. 6.3, that the solutions to the constrained alternative models are $\hat{F}_1 = 1.13F - 0.13F_2$ and $\hat{F}_2 = 1.26F - 0.26F_1$. If these two are rewritten in terms of the solution to the original model, which was $F = 0.64F_1 + 0.36F_2$ (p.94), we obtain $F = 0.88\hat{F}_1 + 0.12F_2$ and $\hat{F} = 0.21F_1 + 0.79\hat{F}_2$ respectively, which are very different answers. This illustrates one of the problems with the simple mixing model, that it is not valid to rewrite the model after the calculation has been performed as quite misleading results may occur.

6.3.4 The generalized mixing model

The generalized mixing model (Le Maitre, 1979) is a logical extension of the simple mixing model, and basically solves the problem of what proportions of two sets of compositions best approximate each other. For example, it can be used in metamorphism to test whether a reaction could be isochemical, by equating the initial reactants and final products; or in metasomatic processes to test models of the type "fresh rock + introduced materials" = "altered rock + removed materials"; or even in igneous petrology to determine how closely two igneous trends approach each other. Greenwood (1967) attempted to solve this problem, which he called the "n-dimensional tie-line problem", for metamorphic reactions, by using linear programming techniques, but ran into difficulties as will be seen in the next section. Most other petrologists faced with this problem have attempted to solve it by using the simple model and arbitrarily choosing one of the phases to be the single $\mathbf{y}$ composition (e.g. Albarède & Provost, 1977) but, as has just been noted, this produces answers which depend upon the choice of the $\mathbf{y}$ composition and is, therefore, clearly unsatisfactory. An alternative approach was taken by Carmichael (1970), who approximated the stoichiometry of metamorphic reactions involving n phases by balancing the reaction for $n-1$ components but, of course, ambiguities arise due to the choice of components. Gray (1973) also briefly mentions balancing metamorphic reactions in his purely mathematical treatment of material balance equations but, unfortunately, gives no examples.

Mathematically the generalized model can be expressed in matrix form as:-

$$c_1\mathbf{y}_1 + c_2\mathbf{y}_2 + \dots + c_m\mathbf{y}_m = b_1\mathbf{x}_1 + b_2\mathbf{x}_2 + \dots + b_n\mathbf{x}_n \tag{6.33}$$

where $\mathbf{y}_1, \dots \mathbf{y}_m, \mathbf{x}_1, \dots \mathbf{x}_n$ are the specified composition vectors and $c_1, \dots c_m, b_1, \dots b_n$ are the proportions to be estimated. To solve this by the least squares method, S, the quantity to be minimized is given by:-

$$S = \sum (c_1y_{1i} + \dots + c_my_{mi} - b_1x_{1i} - \dots - b_nx_{ni})^2 \tag{6.34}$$

However, it should now be obvious that some sort of constraint must be applied to the coefficients, otherwise an exact solution will always be found with S and all the coefficients zero. As we are dealing with compositions, the logical contraint to apply is that the sum of the coefficients on each side of the equation should be 1 (equation 6.26). Admittedly the equations can be solved if the constraint is applied to one side of the equation only, but the symmetrical treatment seems appropriate as there is no reason for choosing one side rather than the other. Geometrically, this reduces the problem to finding the points of closest approach between two hyperplanes, one passing through m and the other through n defined compositions. The value of S is then the square of the shortest distance between the two hyperplanes, and the compositions within the two hyperplanes at the points of closest approach are then given by:-

$$\hat{\mathbf{y}} = c_1\mathbf{y}_1 + c_2\mathbf{y}_2 + \dots + c_m\mathbf{y}_m \tag{6.35}$$

and

$$\hat{\mathbf{x}} = b_1\mathbf{x}_1 + b_2\mathbf{x}_2 + \dots + b_n\mathbf{x}_n \tag{6.36}$$

Applying the constraints to equation 6.34, we obtain the quantity to be minimized S as:-

$$S = \sum (u_i - c_1w_{1i} - \dots - c_sw_{si} - b_1z_{1i} - \dots - b_rz_{ri})^2 \tag{6.37}$$

where $u_i = y_{mi} - x_{ni}$, $w_{ji} = y_{mi} - y_{ji}$, $z_{ji} = x_{ji} - x_{ni}$, $r = n-1$ and $s = m-1$. Again, as this equation is identical in form to equation 6.24, it can be solved in the same manner to obtain the unconstrained values of $c_1, \dots c_s$ and $b_1, \dots b_r$. The constraints are then reimposed when c_m and b_n are calculated using the fact that:-

$$c_m = 1 - c_1 - \ldots - c_s \quad \text{and}$$
$$b_n = 1 - b_1 - \ldots - b_r$$

(6.38)

If either m or n is 1, the generalized mixing model degenerates into the simple constrained mixing model, so that the same computer program may be used for both (Le Maitre, 1981). In fact any computer program that performs multiple regression can be modified to solve the generalized mixing model by first converting the raw x's and y's into the u's, w's and z's of equation 6.37.

In summary, the following points about the generalized mixing model should be noted:-

1. The method calculates the points of closest approach of a $(n - 1)$-flat and a $(m - 1)$-flat, which are defined by n and m compositions, respectively.

2. In general the $(n - 1)$-flat and the $(m - 1)$-flat will not intersect in p-dimensional space as long as $n + m < p + 2$ (relationship 5.2). It is, therefore, unreasonable to expect exact fits between natural metamorphic assemblages (as opposed to idealized mineral assemblages) unless there is a linear relationship between some of the **x** and **y** compositions, e.g. jadeite + quartz = albite or olivine + quartz = hypersthene. However, if the olivine and hypersthene have different Fe/Mg ratios the hyperplanes may not intersect. The important implication of this is that if metamorphic reactions are isochemical, one must expect the number of phases involved in the reactions to be nearly the same as the number of oxides.

3. The square root of the residual sum of squares, S, is the shortest distance between the $(n - 1)$-flat and the $(m - 1)$-flat.

4. If the $(n - 1)$-flat and the $(m - 1)$-flat have a point in common (i.e. $\mathbf{y}_i = \mathbf{x}_j$), an exact solution will be found with $S = 0$, $c_i = b_j = 1$ and $\hat{\mathbf{y}} = \hat{\mathbf{x}} = \mathbf{y}_i = \mathbf{x}_j$.

5. If the $(n - 1)$-flat and the $(m - 1)$-flat each have a point of similar composition (i.e. $\mathbf{y}_i \approx \mathbf{x}_j$), a good fit will be found with $S \approx 0$, $c_i \approx b_j \approx 1$, and with $\hat{\mathbf{y}} \approx \mathbf{y}_i$ and $\hat{\mathbf{x}} \approx \mathbf{x}_j$. This occurs in one of the examples used by Greenwood (1967, Table 5) who comments that the only similarity between the two mineral assemblages is that the chloritoids are of similar composition.

6. If the $(n - 1)$-flat and the $(m - 1)$-flat contain two or more points of similar composition, the two flats become almost parallel so that their points of closest approach may be at extreme compositions, often in the region of negative oxides. This phenomenon is often accompanied by large positive and negative coefficients. This can happen, for example, in attempts to model some of the reactions found in ultramafic nodules such as:-

Clinopyroxene$_1$ + Orthopyroxene$_1$ + Spinel = Clinopyroxene$_2$

+ Orthopyroxene$_2$ + Garnet

7. If one of the flats contains two similar compositions, similar unstable results may be obtained e.g. by using the compositions of the core and rim of a phase that shows only slight compositional zoning.

Numerical example

As a simple example we will calculate the points of closest approach of two linear igneous trends, each trend being defined by two compositions, using a model of the type:-

"basic" comp. A + "acid" comp. A = "basic" comp. B + "acid" comp. B

The two trends used are those of Adak & Kanaga (Coats, 1952) and Crater Lake (Williams, 1942). The end-member compositions used to define the trends (Table 6.5) were estimated using equation 5.19 with the 1st eigenvectors as determined by Le Maitre (1968), and the means. The values of d' in equation 5.19 were chosen so that the SiO_2 values of the compositions correspond to the maximum and minimum SiO_2 values of the actual rocks. With this example equation 6.37 simplifies to:-

$$S = \sum (u_i - c_1 w_{1i} - b_1 z_{1i})^2$$

with $s = r = 1$, so that by analogy with equation 6.25 the normal equations are:-

$$c_1 \sum w_{1i}^2 + b_1 \sum z_{1i} w_{1i} = \sum u_i w_{1i}$$
$$c_1 \sum w_{1i} z_{1i} + b_1 \sum z_{1i}^2 = \sum u_i z_{1i}$$

which, substituting values from Table 6.5, evalute to:-

$$\begin{bmatrix} 248.9218 & -379.7781 \\ -379.7781 & 588.1570 \end{bmatrix} \cdot \begin{bmatrix} c_1 \\ b_1 \end{bmatrix} = \begin{bmatrix} -179.1569 \\ 278.2129 \end{bmatrix}$$

This gives $c_1 = 0.1319$ and $b_1 = 0.5582$, so that using equations 6.38, $c_2 = 1 - 0.1319 = 0.8681$ and $b_2 = 1 - 0.5582 = 0.4418$. These values are given in Table 6.5 as percentages, together with the residual sum of squares (5.86) and the compositions on each trend that are closest to each other, calculated using equations 6.35 and 6.36.

In addition, Table 6.5 also contains data for the Katmai trend (Fenner, 1926, 1950) and the results of similar calculations for the points of closest approach of the Adak & Kanaga and Katmai trends and the Crater Lake and Katmai trends. It is interesting to note that with all three pairs of trends the points of closest approach are in the intermediate compositions. This means that none of the pairs of trends are diverging from a common "parental" magma of basic composition. If they were, the closest points of approach would approximate the basic "parental" magma. The nearest to this situation is where the most basic Katmai lava is very similar to one of the middle range Adak & Kanaga lavas - the distance between the two closest points being 1.40 wt% units. Similarly, it is clear that the trends are not converging on some single "minimum" point on the p-dimensional space liquidus. It is hoped that this might encourage experimental petrologists to continue to examine the effect of additional phases on Petrogeny's Residua system.

TABLE 6.5

Example of the use of the generalized mixing model to find the points of closest approach of pairs of igneous rock trends, using the three approximately linear trends of Adak & Kanaga (Coats, 1952), Crater Lake (Williams, 1942) and Katmai (Fenner, 1926, 1950). The end-member compositions were estimated using equations 5.19 with the 1st eigenvectors, as determined for the trends by Le Maitre (1968), and the means. For further details see text.

Theoretical end-member compositions

	Adak & Kanaga		Crater Lake		Katmai	
	"basic"	"acid"	"basic"	"acid"	"basic"	"acid"
SiO_2	48.00	61.99	51.00	72.00	54.01	75.02
TiO_2	0.89	0.54	1.03	0.37	0.99	0.30
Al_2O_3	19.02	16.85	19.22	15.20	18.27	13.18
Fe_2O_3	4.60	3.91	2.83	1.16	3.87	0.69
FeO	6.16	2.58	5.77	1.33	5.35	1.55
MgO	5.40	2.45	6.59	0.31	4.64	0.41
CaO	11.38	6.28	9.81	2.00	8.90	1.84
Na_2O	3.25	3.72	3.29	5.01	3.43	4.21
K_2O	1.30	1.68	0.46	2.62	0.54	2.80

Estimated compositions of closest approach of pairs of trends

	Adak & Kanaga - Crater Lake		Adak & Kanaga - Katmai		Crater Lake- Katmai	
	Adak & Kanaga	Crater Lake	Adak & Kanaga	Katmai	Crater Lake	Katmai
SiO_2	60.14	60.28	53.79	53.97	61.42	61.53
TiO_2	0.59	0.74	0.75	0.99	0.70	0.74
Al_2O_3	17.14	17.44	18.12	18.28	17.23	16.45
Fe_2O_3	4.00	2.09	4.31	3.88	2.00	2.73
FeO	3.05	3.81	4.68	5.36	3.57	3.99
MgO	2.84	3.82	4.18	4.65	3.47	3.12
CaO	6.95	6.36	9.27	8.91	5.93	6.37
Na_2O	3.66	4.05	3.44	3.43	4.14	3.71
K_2O	1.63	1.41	1.46	0.54	1.53	1.35

Coefficients for generalized mixing model expressed as percentages

% "basic"	13.19	55.82	58.64	100.22	50.42	64.24
% "acid"	86.81	44.18	41.36	-0.22	49.58	35.76

Residual sum of squares, S^2

	5.8562		1.9670		1.8653	

Shortest distance between trends, S

	2.42		1.40		1.36	

Value of 100a in equation 6.45, for which $S = 1.96 \, SD_S$

	8.76		3.32		2.77	

6.3.5 *Solutions using linear programming*

Wright & Doherty (1970) and more recently Banks (1979) have advocated the use of linear programming methods to solve the simple mixing models, mainly on the grounds that it is easy to impose the constraints on the solutions that a) none of the proportions (coefficients) should be negative and that b) they should sum to 1. The mathematics of linear programming will not be described here as it is immaterial to the discussion, but interested readers may refer to Kim (1971) or Trustrum (1971) or to the clear and concise description given by Harbaugh & Bonham-Carter (1970). However, the relevance and importance of these two constraints has already been discussed in section 6.3.2, with the conclusion that the second constraint is the really important one. There would appear to be no advantage, therefore, in using linear programming, as the second constraint, unlike the first, is easily incorporated into a least squares solution.

An unusual feature of the method of Wright & Doherty (1970) is that they use linear programming only to select the best combination of compositions from the many possibilities which give non-negative values of b_j, as their final answer is obtained using the least squares method. Their method also differs from the simple model by allowing the oxides to be weighted.

Prior to the least squares solution of the simple mixing model, Greenwood (1967) attempted to solve a generalized mixing model involving two metamorphic assemblages by using linear programming methods, with the first of the two constraints mentioned above. He found, however, that with natural assemblages it was difficult to obtain a solution to the problem (no intersection of the hyperplanes) without introducing analytical errors to all the values. In view of what has previously been said about the intersection of hyperplanes in p-dimensional space (equation 5.2), the reason for the lack of solutions should now be obvious. Later, Greenwood (1968) used regression methods to solve the same problem in a way equivalent to the unconstrained simple mixing model.

6.3.6 *Weighted mixing models*

In all the models discussed so far, it has been assumed that the analyses only contain major elements. If, however, it is desired to use both major and trace element data in the model, then weighted mixing models should be used to take into account the differing magnitudes of the numbers and of the errors in their measurements. Reid et al. (1973) go so far as to assert that all simple petrological mixing calculations should use weighted models, on the grounds that if weighting factors are not used, equal absolute errors are assumed for all oxides.

Although the theory of the solution of weighted simple mixing models is somewhat lengthy (e.g. see Gray, 1973 and Reid et al., 1973), the final outcome is only a minor modification of the unweighted model. All that is required is to transform the raw data, by multiplying each of the ith oxide values by a weighting factor $1/f_i$, where f_i is an *a priori* estimate of the error in the determination of the ith oxide. The method assumes that there is no correlation between the errors in the oxides, which is reasonable if the oxides have been independently determined by one of the instrumental techniques, such as x-ray fluorescence or electron microprobe analysis. Another point to note is that the f_i do not have to be absolute values, but can be given in relative terms. For example, if it is felt that the absolute error in MgO is twice that for FeO, then the values of f_{MgO} and f_{FeO} could be given as 2 and 1, or 1 and 0.5, or 10 and 5, as long as the ratio between them remains constant.

The application of weighting to the generalized mixing model, however, poses the problem of having to assume that each oxide has a constant standard deviation which means, for example, that the absolute standard deviation of MgO in an olivine would have to be assumed to be the same as that in a magnetite. This is contrary to the usually accepted idea of the behaviour of the error of analytical results, whereby the standard deviation of an oxide is assumed to be of the form:-

$$SD_x = a + bx \qquad (6.39)$$

where a is a constant to take into account the lower error threshold, b is a proportionality, and x is the concentration of the oxide (Albarède & Provost, 1977).

6.3.7 *Problem of errors associated with petrological mixing models*

So far, all published methods of dealing with the errors associated with petrological mixing have been confined to the simple models. For example, Wright & Doherty (1970) adopt an empirical iterative procedure to test "the sensitivity of the residuals to small finite changes in the calculated solution values". This is done by calculating the maximum changes that can be made to the proportions before either the individual or average residuals exceed arbitrarily set limits. These maximum changes are then regarded as the "errors" in the estimated proportions. However, they give no method to test if the overall fit, as measured by the residual sum of squares, is significant.

Reid et al. (1973), on the other hand, utilize the least-squares error theory, and obtain solutions which are dependent on *a priori* estimates of the errors of the individual oxides in the single composition to be estimated, i.e. the y_i of equation 6.22, but again do not attempt to judge the overall fit.

Banks (1979), however, uses linear programming methods to obtain 95% confidence limits for the proportions and suggests that the overall fit can be tested for significance by treating the standardized residual sum of squares as a chi-squared statistic with p d.f. Using the terminology of equation 6.24, this can be expressed as:-

$$\chi^2 = \sum \left[(y_i - y_i) / SDy_i^{\ 2} \right]$$ (6.40)

where SDy_i is the standard deviation of y_i. However, this is unsatisfactory as the chi-squared statistic assumes that each term of the summation is independent, which is not the case with the closed data of chemical analyses. In order to implement a solution by means of linear programming, he also suggests a measure based on the sum of the absolute values of the individual misfits, which is also suspect for the same reason as the chi-squared statistic. Even if the violation of these assumptions is trivial, difficulties would still arise in applying both this and the previous method to the generalized mixing model, as both methods assume that the standard deviation of an oxide is constant. This is the same problem that was encountered with weighting generalized mixing models. The problem is further complicated in the generalized mixing model by the fact that any distinction between "independent" and "dependent" variables is entirely artificial, due to the symmetry of the equation.

However, it is doubtful if any method of error analysis is valid which does not take into account the experimental error in each oxide determination, as petrological mixing models are not statistical problems of the type where the errors can be calculated from the data itself. For example, in multiple regression a model is set up involving a random variable (equation 6.1) to take into account the fact that the proposed model is only approximate, and that the independent variable will never explain the dependent variable *exactly*. In petrological mixing, however, one is seeking to establish an exact relationship as, for example, a rock composition must be expressible exactly in terms of its constituent minerals. If it cannot be expressed exactly, the discrepancies must be due either to an incorrect choice of model or to errors of measurement - they are never due to a random source in the model.

Viewed in this light the problem has a simple geometric interpretation. Given that the generalized petrological mixing model determines the shortest distance between two hyperplanes defined by n and m compositions, it can be seen that "slop" will be introduced into the location of each hyperplane if every oxide is subject to an error of measurement, which may be estimated or assumed. In practice some of these errors may be zero as, for example, when theoretical compositions, such as 100% SiO_2 for quartz, are used in the model. The problem of testing the "goodness-of-fit" is then reduced to the problem of determining the error in the estimated shortest distance between the two hyperplanes, when all the compositions defining the hyperplanes are subject to possible errors. This error can be estimated by applying the principle of the propagation of errors, as described in Chapter 11, to the generalized mixing model of equation 6.34, from which we obtain for use in equation 11.3:-

$$\frac{\partial S}{\partial y_{ij}} = 2e_i c_j \quad \text{and} \quad \frac{\partial S}{\partial x_{ki}} = -2e_i b_k \tag{6.41}$$

where e_i is the misfit of the ith oxide given by:-

$$e_i = c_1 y_{1i} + \dots + c_m y_{mi} - b_1 x_{1i} - \dots - b_n x_{ni} \tag{6.42}$$

which by substitution into equation 6.41 gives:-

$$\text{var } S = \sum_{i=1}^{p} \sum_{j=1}^{m} (2e_i c_j)^2 \text{ var } y_{ji} + \sum_{i=1}^{p} \sum_{k=1}^{n} (-2e_i b_k)^2 \text{ var } x_{ki} \tag{6.43}$$

If the standard deviations of the individual oxides are known, they should be used. In practice, however, they are often unknown, in which case the following strategy can be used to provide a general insight into the signficiance of the fit. Assuming that all the errors are of the form of equation 6.39, but with a = b, the standard devation of each oxide can be written as:-

$$SD_x = a(1 + x) \tag{6.44}$$

The variance terms of equation 6.43 can then be replaced to give, with some slight rearrangement:-

$$\text{var } S = 4a^2 \sum_{i=1}^{p} e_i^2 \left[\sum_{j=1}^{m} (c_j)^2 (1 + y_{ji})^2 + \sum_{k=1}^{n} (b_k)^2 (1 + x_{ki})^2 \right] \tag{6.45}$$

Using this equation, it is then possible to calculate the value of a at which the shortest distance between the two hyperplanes is equal to 1.96 times its standard deviation. This is equivalent to testing the null hypothesis that, at the 95% confidence level, the shortest distance is zero. These values of a, expressed as percentages, are given for the two examples of mixing models used in Tables 6.4 and 6.5. As can be seen in Table 6.4, the mode of the dacite can be regarded as a perfect fit, as it is unlikely that the overall relative errors in the oxide determinations are less than 0.5%! In Table 6.5, however, the overall errors would have to be as large as 8.76% for the fit between Adak & Kanaga and Crater Lake to be considered perfect, which seems unlikely. Judging the fit of the other two trends in Table 6.5 is much more subjective, and depends critically upon the assumptions made about the errors.

Another method, which also utilizes *a priori* estimates of the standard deviations in all the oxide determinations, is that given by Albarède & Provost (1977). Unfortunately the method only deals with simple mixing models, but it has the advantage that a computer program is included. The method is mathematically complicated and involves a maximum-likelihood solution to maximize a standardized chi-square function, which then provides information on the significance of the overall fit, as well as the errors associated with the proportions.

CHAPTER 7

PRINCIPAL COMPONENTS ANALYSIS

7.1 INTRODUCTION

Over the decades, much effort in petrological research has been put into the construction of *variation diagrams*, usually to illustrate the relationships within a series of rock analyses. Choosing the two or three variables to be plotted from among the many possibilities, however, has always been something of a problem. In the end the choice has usually been dictated either by simplicity, or by what has been done before, or by some preconcieved idea of how the data behave. Many of these plots suffer from the disadvantage that much of the raw data is not used, resulting in a loss of information in the diagram. For somewhat similar reasons, criticism has also been levelled at the common use in igneous petrology of certain triangular diagrams, such as the AFM (Wright, 1974). In an attempt to overcome these problems, some petrologists have used multiple plots. A classic example of this is the Harker diagram, in which each oxide is plotted against SiO_2, but although this diagram contains all the information, it is difficult to use especially when making comparisons. An alternative solution is to use the method of principal component analysis, as advocated by Le Maitre (1968) and more recently by Till & Colley (1973).

Principal components analysis is probably one of the most valuable of all the multivariate statistical techniques and can be useful in two major ways. Firstly, by means of a rigid rotation of axes, it can be used to transform the original set of variables into a new set of variables, called principal component coordinates (Gnanadesikan, 1977), which display a decreasing amount of variance, and are uncorrelated with each other. When plotted against each other, these coordinates provide the optimum way of viewing the data. Secondly, if the first few sets of principal component coordinates account for a high proportion of the variance, the remainder may be discarded thereby reducing the number of variables that need to be considered. This also gives an insight into the number of processes that could be involved in causing the variation within the original data.

As Marriott (1974, p.18) has stated "the method is perfectly general; it involves no assumption about the original variables, no hypothesis that can be tested, no underlying model. It is simply a different, and possibly more convenient, way of expressing the same set of results". In this respect it is markedly different from factor analysis, with which it is often equated. This point is discussed further in the next chapter.

7.2 GEOMETRIC INTERPRETATION

A simple way to understand the method of principal components analysis is to consider it in terms of the geometry of p-dimensional space, as shown diagramatically with a 3-dimensional example in Fig. 7.1. The method first defines the direction MV_1, in which there is the maximum amount of "spread" of the data; M being the mean of all the data points. This direction is known as the *1st eigenvector* (or *characteristic vector* or *latent vector*) and can be thought of as the "best-fit straight line" in p-dimensional space (see also p.55). The original data can then be transformed into the 1st set of *principal component coordinates* (also known as *principal components scores*), by using equation 5.20 to project the data onto the 1st eigenvector. The variance of these coordinates is the *1st eigenvalue* (or *characteristic root* or *latent root*) and is a measure of "spread" in the direction of the eigenvector.

The method then defines a 2nd eigenvector, MV_2, which has the maximum amount of "spread" at right angles to the 1st eigenvector; a 3rd eigenvector, MV_3, which has the maximum amount of "spread" at right angles to the two previous eigenvectors, and so on. In general, in p-dimensional space this process goes on until p eigenvectors, and their associated eigenvalues, have been defined. By definition, the magnitudes of the eigenvalues are in

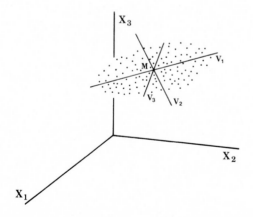

Fig. 7.1. Diagram to illustrate the geometric interpretation of principal components analysis. In the 3-dimensional space defined by the axes OX_1, OX_2 and OX_3, the cluster of data points has three eigenvectors, MV_1, MV_2 and MV_3, where M is the mean of the data. MV_1 is the 1st eigenvector and defines the direction of maximum "spread" of the data, i.e. the variance of the data projected onto MV_1 is a maximum. The 2nd eigenvector, MV_2, defines the direction of maximum "spread" at right angles to MV_1; and the 3rd eigenvector, MV_3, defines the direction of maximum "spread" at right anlges to both MV_2 and MV_3.

decreasing order. If the data belong to a *multivariate normal distribution*, the eigenvectors also define the principal axes of the ellipsoids of equal probability, as shown in Fig. 4.1.

As a result of the way in which the eigenvectors are defined, it follows that by plotting the first two or three sets of principal component coordinates against each other, simple plots can be produced which will contain a maximum amount of information. The amount of variance contained in the plot will depend upon the values of the corresponding eigenvalues but, by definition, no other projection in the same p-dimensional space, and on the same number of axes, will contain more. This useful property is often referred to as "reducing the dimensionality of the problem" and is one of the major reasons for using principal components analysis. Another way of looking at the method is that it is essentially a way of redistributing the variances.

The prime importance of principal components analysis for the production of petrological variation diagrams, should now be self-evident. By examining the eigenvalues, it is possible to deduce something about the overall shape of the distribution of the data in p-dimensional space and about the number of possible processes that could have been responsible for the variation in the data. For example, if the 1st eigenvalue is large compared with the others, it is possible that one major petrological process caused the variation in the data – the assumption made in the second example in section 5.4. On the other hand, if both the 1st and 2nd eigenvalues are large, it is probable that two processes operated, and so on. However, as principal components analysis is a type of linear fitting, a well-defined curved trend in p-dimensional space may be indistinguishable from an ellipsoidal distribution as illustrated in Fig. 7.2. It is imperative, therefore, that the data are always plotted and examined before deciding upon the cause of variability. This strategy, in fact, should be adopted before any multivariate analysis is performed.

7.3 SCALING THE VARIABLES

It must be emphasized that there are two common methods of performing principal components analysis, which depend upon whether or not the data have been standardized. The two sets of eigenvectors and eigenvalues will be different and there is no simple way of calculating one set from the other. The

Fig. 7.2. Diagram to illustrate how three different distributions could produce similar sets of eigenvalues, as the ratios of the variances along the two eigenvectors are approximately the same for each distribution.

petrologist must, therefore, decide which method to use for a particular problem and, for the results to be of use to other researchers, must state which method was used.

In one method, the raw data are unscaled and the eigenvectors and eigenvalues are extracted from the symmetrical *variance-covariance* or *dispersion matrix*:-

$$\mathbf{D} = \begin{bmatrix} \text{var } x_1 & \text{cov } x_1x_2 & \text{cov } x_1x_3 & \cdots & \text{cov } x_1x_p \\ \text{cov } x_2x_1 & \text{var } x_2 & \text{cov } x_2x_3 & \cdots & \text{cov } x_2x_p \\ \text{cov } x_3x_1 & \text{cov } x_3x_2 & \text{var } x_3 & \cdots & \text{cov } x_3x_p \\ \cdots & \cdots & \cdots & \cdots & \cdots \\ \text{cov } x_px_1 & \text{cov } x_px_2 & \text{cov } x_px_3 & \cdots & \text{var } x_p \end{bmatrix} \tag{7.1}$$

where $\text{var } x_i$ and $\text{cov } x_ix_j$ are the variance and covariance, respectively, as defined by equations 2.4 and 4.3. Alternatively, the variance-covariance matrix may be calculated by using equations 8.12 and 8.14. As the data are unscaled, this method should only be used when all the variables are measured in the same units, such as wt% or ppm, otherwise, for example, a variation of 450 to 650 ppm (200 units) in one trace element will be given far more weight than a variation of 45% to 65% (20 units) in a major oxide, which may not be desired or reasonable.

In the other method the raw data are standardized, so that each variable has unit variance and zero mean. In this case the eigenvectors and eigenvalues are extracted from the *correlation matrix*, **R**, which can be calculated in several ways. One common method is to divide each term of the variance-covariance matrix, **D**, by the square root of the product of the diagonal terms of the row and column in which it lies. This is simply using equation 4.4 and gives the correlation matrix:-

$$\mathbf{R} = \begin{bmatrix} 1 & r_{12} & r_{13} & \cdots & r_{1p} \\ r_{21} & 1 & r_{23} & \cdots & r_{2p} \\ r_{31} & r_{32} & 1 & \cdots & r_{3p} \\ \cdots & \cdots & \cdots & \cdots & \cdots \\ r_{p1} & r_{p2} & r_{p3} & \cdots & 1 \end{bmatrix} \tag{7.2}$$

where r_{ij} is the correlation coefficient between the ith and jth variable (section 4.2). A more time consuming method is to first standardize all the variables, by using equation 1.2, and then to calculate the variance-covariance matrix as above, which will then be identical to the correlation matrix. Another method is to use equations 8.13 and 8.14. By using the correlation matrix, all the variables are given equal weight so that, for example, a variation of 40% in one major oxide may be given as much importance as a variation of 0.4% in another. This is the method commonly used by workers in the behavioural and life sciences, as their variables are often measured on completely different scales. Petrologists should also be aware that some of the statistical computer packages automatically use the correlation matrix for principal components analysis.

Geometrically the difference between the two methods is simple. When the data are standardized, and the correlation matrix is used, those axes corresponding to variables with a small variance are effectively stretched, while those with a large variance are compressed. Experience with both methods suggests that standardization usually leads to a 1st eigenvector in which a large number of the terms are of similar magnitude. Also the percentage amount of variance accounted for by the first two eigenvalues is invariably smaller than if the principal components analysis had been performed with the variance-covariance matrix. As a guide to the choice of method, the following suggestions are made:-

1. *If all the variables are measured in the same units* e.g. wt% or ppm, use the raw data and extract the eigenvectors and eigenvalues from the variance-covariance matrix, as recommended by many authors (e.g. Davis, 1973; Gnanadesikan, 1977; Marriott, 1974; Morrison, 1967). However, Till & Colley (1973), have advocated using standardized variables for petrological data on the ground that "the alkalies and titanium begin to show their true importance" - this approach, however, does appear to be prejudging the answer. It is interesting to note that some authors who have used principal components analysis to study the major element chemistry of mineral groups (e.g. Saxena, 1969a, 1969b, 1969c; Saxena & Ekström, 1970; Saxena & Walter, 1974; Saxena et al., 1977), have also used the correlation matrix and, therefore, standardized variables. This seems rather illogical, as cations in minerals are not free to vary independently as they are strictly controlled by the crystal structure and replacement is essentially on a one-for-one basis, which surely means that the variables should left as they are. The first numerical example will illustrate this point.

2. *If the variables are measured in different units* the usual strategy is to standardize the variables by using the correlation matrix. An example of this is found in Hawkins (1974) who uses principal components analysis on the major elements and cell parameters of some zeolites. Standardization is also often used when both wt% and ppm are used together. However, in this case there is an alternative solution, which is to convert all the variables to the same scale (either wt% or ppm) and then to use the logarithms of the values as the raw data. This approach has the attraction that changes of the same order of magnitude will be given the same weight, e.g. a change from 400 ppm to 4000 ppm in a trace element will be given the same weight as a change from 1% to 10% in a major oxide.

Numerical example

Consider the problem of preparing a variation diagram to illustrate the "trend" shown by the analyses of Table 7.1. A simplistic approach would be to note the tendency for high values of FeO to be associated with low values of MgO and, as both of these variables have a large variance, to plot FeO against MgO as shown in Fig. 7.3a.

TABLE 7.1

Some simplified pyroxene analyses taken from Deer et al. (1966, Table 13). The original analysis numbers have been retained but each analysis has been recalculated to sum to 100%. $R_2O_3 = TiO_2 + Al_2O_3 + Fe_2O_3 + Cr_2O_3$; FeO = FeO + MnO; MgO = MgO + NiO.

Analysis	SiO$_2$	R$_2$O$_3$	FeO	MgO	CaO
1	57.75	1.87	3.65	36.50	0.23
2	49.97	4.20	28.64	15.75	1.44
3	46.48	1.33	47.21	3.53	1.45
4	54.99	4.70	1.59	17.27	21.45
5	48.72	1.89	26.85	1.07	21.47
6	52.85	5.02	5.71	16.48	19.94
7	47.58	5.67	21.73	7.42	17.60
8	50.11	4.67	18.90	16.33	9.99
9	49.79	8.79	26.33	6.99	8.10
10	53.06	1.73	17.52	23.61	4.08
11	50.51	3.53	29.21	12.89	3.86
Means	51.07	3.95	20.67	14.35	9.96
S.D.	3.32	2.22	13.43	9.97	8.59

However, a better approach is to use principal components analysis and, as all the data are measured in the same units (wt%), to extract the eigenvectors from the variance-covariance matrix. This gives the results shown in Table 7.2, which indicate that the first two eigenvalues account for over 98% of the variance, so that the first two principal components contain virtually all of the variation in the original data. Inspection of the 1st eigenvector indicates that FeO tends to vary inversely with MgO as both of these terms are of opposite sign and reasonably large. Similarly, the 2nd eigenvector indicates that CaO varies inversely with MgO and FeO. Hence, having assumed absolutely nothing about the crystal chemistry of pyroxenes, for that is what the analyses of Table 7.1 are, principal components analysis has revealed that the majority of the variation is related to changes in CaO, FeO and MgO. The 3rd eigenvector shows that R$_2$O$_3$ varies inversely with CaO, MgO and FeO which is, of course, a measure of the compositional departure from the pyroxene quadrilateral (Di-Hd-En-Fs). If this eigenvector was not interpretable in terms of the known chemistry of pyroxenes, it would probably be ascribed to "noise" due to the small size of its eigenvalue. As the data of Table 7.1 have been recalculated to 100% the variance-covariance matrix is *singular* and the 5th eigenvalue is, therefore, zero. The 5th eigenvector then defines the normal to the plane of constant composition in the 5-dimensional space, so that each term is equal to $1/\sqrt{5} = 0.447$ (equation 7.10).

A plot of the pyroxenes projected onto the first two eigenvectors, using the principal component coordinates of Table 7.2, is shown in Fig. 7.3c, in which the pyroxene quadrilateral has also been projected. The similarity between this plot and the conventional pyroxene plot shown in Fig. 7.3b,

TABLE 7.2

Results of principal components analysis of the pyroxene data of Table 7.1, with the eigenvectors and eigenvalues extracted from the variance-covariance matrix.

Variance-covariance matrix

	SiO_2	R_2O_3	FeO	MgO	CaO
SiO_2	10.9973	-0.9531	-38.0521	29.3483	-1.3404
R_2O_3	-0.9531	4.9211	-4.3839	-5.1608	5.5766
FeO	-38.0521	-4.3839	180.2756	-91.6790	-46.1605
MgO	29.3483	-5.1608	-91.6790	99.3070	-31.8154
CaO	-1.3404	5.5766	-46.1605	-31.8154	73.7396

Eigenvectors extracted from the variance-covariance matrix

	1st	2nd	3rd	4th	5th
SiO_2	0.192	0.082	0.079	-0.866	0.447
R_2O_3	0.006	-0.079	-0.879	0.145	0.447
FeO	-0.830	0.239	0.215	0.089	0.447
MgO	0.509	0.550	0.250	0.419	0.447
CaO	0.123	-0.792	0.335	0.213	0.447
Eigenvalues	252.16	110.46	5.42	1.20	0.00
As %	68.29	29.92	1.47	0.32	0.00
Accum. %	68.29	98.21	99.68	100.00	100.00

Principal component coordinates calculated using equation 7.16

Analysis	1st	2nd	3rd	4th	5th
1	26.69	25.38	12.92	-34.07	44.72
2	-5.95	18.15	10.83	-33.21	44.72
3	-28.27	15.79	14.01	-34.07	44.72
4	20.70	-2.95	12.06	-34.99	44.72
5	-9.73	-6.14	15.42	-34.50	44.72
6	16.28	-1.40	11.79	-33.38	44.72
7	-2.92	-1.20	11.20	-31.59	44.72
8	3.51	9.34	11.35	-32.06	44.72
9	-7.68	7.12	6.33	-34.85	44.72
10	8.18	18.18	13.71	-33.38	44.72
11	-7.48	14.89	11.68	-34.41	44.72

Theoretical end-member pyroxenes

Analysis	1st	2nd	3rd	4th	5th
Diopside	23.32	-5.69	17.72	-34.73	44.72
Hedenbergite	-11.95	-6.99	17.62	-34.55	44.72
Enstatite	31.94	27.03	14.78	-34.99	44.72
Ferrosilite	-36.45	16.76	15.30	-34.59	44.72

TABLE 7.3

Results of principal components analysis of the pyroxene data of Table 7.1, with the eigenvectors and eigenvalues extracted from the correlation matrix.

Correlation matrix

	SiO_2	R_2O_3	FeO	MgO	CaO
SiO_2	1.0000	-0.1296	-0.8546	0.8881	-0.0471
R_2O_3	-0.1296	1.0000	-0.1472	-0.2335	0.2927
FeO	-0.8546	-0.1472	1.0000	-0.6852	-0.4004
MgO	0.8881	-0.2335	-0.6852	1.0000	-0.3718
CaO	-0.0471	0.2927	-0.4004	-0.3718	1.0000

Eigenvectors extracted from the correlation matrix

	1st	2nd	3rd	4th	5th
SiO_2	-0.605	0.007	0.003	0.778	0.173
R_2O_3	0.087	0.557	0.817	0.035	0.115
FeO	0.542	-0.371	0.085	0.270	0.699
MgO	-0.576	-0.234	0.171	-0.562	0.519
CaO	0.031	0.705	-0.544	-0.079	0.447
Eigenvalues	2.63	1.57	0.74	0.06	0.00
As %	52.67	31.30	14.81	1.22	0.00
Accum. %	52.67	83.97	98.78	100.00	100.00

Principal component coordinates calculated using equations 7.17 or 7.18

Analysis	1st	2nd	3rd	4th	5th
1	-3.30	-1.36	0.13	0.03	0.00
2	0.42	-0.89	0.71	-0.09	0.00
3	2.40	-1.84	-0.45	0.10	0.00
4	-1.58	1.60	-0.52	0.28	0.00
5	1.41	0.56	-1.68	0.18	0.00
6	-0.97	1.46	-0.29	-0.08	0.00
7	1.18	1.19	0.04	-0.45	0.00
8	0.02	0.18	0.29	-0.36	0.00
9	1.07	1.08	1.81	0.32	0.00
10	-1.13	-1.17	-0.30	-0.10	0.00
11	0.49	-0.81	0.26	0.17	0.00
Theoretical end-member pyroxenes					
Diopside	-1.98	0.80	-2.52	0.17	0.00
Hedenbergite	1.54	0.15	-2.45	0.18	0.00
Enstatite	-4.12	-1.83	-0.50	0.22	0.00
Ferrosilite	3.01	-2.42	-0.86	0.22	0.00

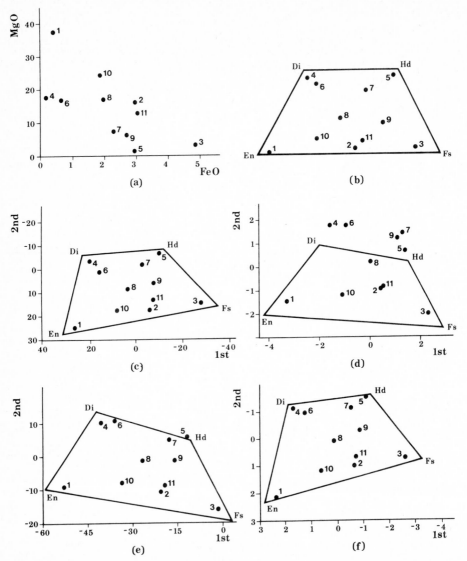

Fig. 7.3. Various plots of the pyroxene data of Table 7.1 to illustrate the differing spatial arrangements that can result from projections derived from the correct and incorrect application of the two different methods of principal components analysis:- (a) an intuitive "variation diagram" of MgO versus FeO; (b) a conventional mineralogical plot into the Di–Hd–En–Fs quadrilateral; (c) a projection into the space of the first two eigenvectors derived from the variance–covariance matrix, formed by plotting the first two sets of principal component coordinates from Table 7.2 against each other; (d) a projection into the space of the first two eigenvectors derived from the correlation matrix, formed by plotting the first two sets of principal component coordinates from Table 7.3 against each other. Note how several of the pyroxenes now lie outside the pyroxene quadrilateral; (e) a projection using the incorrectly calculated "principal component coordinates" of Method A from Table 7.4; (f) a projection using the incorrectly calculated "principal component coordinates" of Method B from Table 7.4.

constructed with a knowledge of the crystal chemistry of pyroxenes, is self-evident.

If the eigenvectors are extracted from the correlation matrix, as is commonly done, different results are obtained as shown in Table 7.3. One obvious difference is that the first two eigenvalues now account for less than 84% of the total variance, instead of 98% as before, and that the 3rd eigenvalue is of considerable magnitude. The interpretation of the eigenvectors is also different, with the 1st showing that FeO varies inversly with SiO_2 and MgO, the 2nd showing that R_2O_3 and CaO vary inversely with FeO and MgO, while the 3rd shows an inverse relationship between R_2O_3 and CaO. In the light of the known crystal chemistry of the pyroxenes, these eigenvectors are more difficult to explain than those extracted from the variance-covariance matrix.

A plot of the pyroxenes projected onto these first two eigenvectors, using the principal component coordinates of Table 7.3, is shown in Fig. 7.3d. As can be seen, several of the analyses now plot outside the pyroxene quadrilateral in positions that do not reflect their known mineralogy, a situation which, to most people, would be untenable. It is for this reason that the author believes it is unrealistic to extract the eigenvectors from a correlation matrix if all the variables have been measured in the same units (c.f. Till & Colley, 1973).

7.4 BASIC MATHEMATICS

The detailed mathematical theory behind the extraction of eigenvectors and eigenvalues is not relevant to this book, as the work is nearly always performed by standard computer subroutines. If required, detailed accounts may be found in many texts, including Cooley & Lohnes (1971), Morrison (1967), and Tatsuoka (1971). However, it is useful to understand some of the mathematical properties of the matrices used in principal components analysis and their eigenvectors and eigenvalues.

If **A** is a symmetrical pxp matrix, with terms a_{ij}, from which the eigenvectors and eigenvalues are extracted, v_i is the ith eigenvector whose terms are v_{ij}, and λ_i is the ith eigenvalue, it can be shown that:-
1. The eigenstructure of **A** is given by the matrix expression:-

$$(\mathbf{A} - \lambda_i \mathbf{I})\mathbf{v}_i = 0 \tag{7.3}$$

where **I** is an identity matrix. Alternatively this may be written as:-

$$\mathbf{A}\mathbf{v}_i = \mathbf{v}_i \lambda_i \tag{7.4}$$

or in full matrix form as:-

$$\mathbf{A}\mathbf{V} = \mathbf{V}\mathbf{\Lambda} \tag{7.5}$$

where **V** is a pxp matrix of all the eigenvectors and **Λ** (Greek letter capital lambda) is a pxp diagonal matrix with the eigenvalues as its diagonal terms. Postmultiplying both sides by $\mathbf{V}^*$, the transpose of **V**, then gives the relationship that:-

$$\mathbf{A} = \mathbf{V} \mathbf{\Lambda} \mathbf{V}^* \tag{7.6}$$

2. The sum of the eigenvalues is equal to the sum of the diagonal terms of **A** or:-

$$\sum \lambda_i = \sum a_{ii} \tag{7.7}$$

This enables eigenvalues to be expressed as percentages without all of them having to be determined. This can be seen from Table 7.2, in which the sum of the eigenvalues and the sum of the diagonal terms of the variance-covariance matrix are both equal to 369.24, and Table 7.3 for which the corresponding value is 5. This relationship also holds for certain non-symmetrical matrices (see section 9.3).

3. By definition:- $\lambda_1 \geqslant \lambda_2 \ .. \ .. \geqslant \lambda_p$ (7.8)

4. If **A** is a singular matrix (e.g. if the original data are closed, such as major element data recalculated to sum to 100%), then λ_p, the pth eigenvalue, will be zero (see Tables 7.2 and 7.3), assuming that there are no rounding-off errors in the calculation. In practice the value should be extremely small and with some computer programs may even be negative. If the raw data are not exactly closed, then the matrix **A** will be nearly singular (ill-conditioned) and λ_p will have a small positive value.

5. From equation 7.4 it follows that the terms of $\mathbf{v}_i$ are not unique, for if $\mathbf{v}_i$ is a valid eigenvector, so is $c\mathbf{v}_i$, where c is any constant. It is for this reason that the eigenvectors are normalized and given in direction cosine form so that we have:-

$$\sum_{j=1}^{p} v_{ij}^2 = 1.0 \tag{7.9}$$

For example, the sum of squares of the terms of the 2nd eigenvector in Table 7.2 is:-

$$0.082^2 + 0.079^2 + 0.239^2 + 0.550^2 + 0.792^2 = 0.999850$$

which, within rounding-off errors, is 1. For the interpretation of such vectors see section 5.3.

6. By definition, the eigenvectors are orthogonal, so that using equation 5.10 to calculate the cosine of the angle between two eigenvectors we have:-

$$\sum_{k=1}^{p} v_{ik} v_{jk} = 0.0 \qquad \text{for } i \neq j \tag{7.10}$$

For example, in Table 7.3 the cosine of the angle between the 1st and 3rd eigenvectors is given by:-

$$(-0.605 \times 0.003) + (0.087 \times 0.817) + ... + (0.031 \times -0.544) = -0.000026$$

which, within rounding-off errors, is zero.

7. If **A** is a *singular variance-covariance matrix* (as in Table 7.2), it can be shown that:-

$$\sum_{j=1}^{p} a_{ij} = 0.0 \tag{7.11}$$

which is a restatement of equation 4.12, and that:-

$$\sum_{j=1}^{p} v_{ij} = 0.0 \quad \text{for } i \neq p \tag{7.12}$$

It also follows that each term of the pth eigenvector is:-

$$v_{pj} = 1/\sqrt{p} \tag{7.13}$$

i.e. the eigenvector defines the normal to the plane of constant composition (section 5.1).

8. If **A** is a *singular correlation matrix* (as in Table 7.3), each term of the pth eigenvector is proportional to SDx_i, the standard deviation of its corresponding variable, so that:-

$$v_{pi} = c \, SDx_i \tag{7.14}$$

where c is a constant. This can be seen in Table 7.3 where, if the terms of the 5th eigenvector are each multiplied by 19.21, they are equal to the standard deviations of the corresponding variables as given in Table 7.1.

9. If **A** is a *variance-covariance matrix*, z_{ij}, the jth principal component coordinate of the ith set of data $x_{i1}, \dots \dots x_{ip}$, should be calculated using the equation:-

$$z_{ij} = \sum_{j=1}^{p} v_{jk}(x_{ik} - \bar{x}_k) \tag{7.15}$$

which is based on equation 5.20. With this equation the value of z_{ij} is the coordinate of the ith data point when it is projected onto a direction parallel to the jth eigenvector and with the origin at the location of the mean of all the data points. However, if we are prepared to tolerate a shift of origin, we can use the simpler equation:-

$$z_{ij} = \sum_{j=1}^{p} v_{jk}x_{ik} \tag{7.16}$$

which gives the coordinate of the ith data point when it is projected onto a direction parallel to the jth eigenvector, but with the same origin as that of the data. For example, in Table 7.2, z_{23}, the 3rd principal component coordinate of the 2nd pyroxene analysis in Table 7.1, is given by:-

$$z_{23} = (0.079 \times 49.97) + (-0.879 \times 4.20) + ... + (0.335 \times 1.44) = 10.83$$

Expansion of equation 7.15 shows that the only difference between it and equation 7.16 is the presence of a constant term for each eigenvector. In practice equation 7.16 is the expression that is normally used for calculating principal component coordinates. However, *neither* of these expressions should be used if the eigenvectors have been calculated from a correlation matrix, as this is equivalent to an *oblique rotation of axes*.

10. If **A** is a *correlation matrix*, then the original variables must be standardized, by using equation 1.2, before using equation 5.20, so that we must calculate the principal component coordinates using the expression:-

$$z_{ij} = \sum_{j=1}^{p} v_{jk} (x_{ik} - \bar{x}_k) / SDx_k \qquad (7.17)$$

where $\bar{x}_k$ and SDx_k are the mean and standard deviation of the kth variable, respectively. As can be seen, p subtractions, divisions, multiplications and additions have to be performed to calculate one principal component coordinate. However, by rewriting the equation as:-

$$z_{ij} = \sum_{j=1}^{p} v_{jk}'x_{ik} - c_j \qquad (7.18)$$

the computational effort can be nearly halved, as $v_{jk}' = v_{jk}/SDx_k$ and $c_j = \sum v_{jk}'\bar{x}_k$ need only be calculated once. For example, using equation 7.18 to determine z_{23}, the 3rd principal component coordinate in Table 7.3, of the 2nd pyroxene analysis in Table 7.1, we obtain:-

$$z_{23} = (0.000904 \times 49.97) + ... + (-0.063329 \times 1.44) - 1.246 = 0.71$$

11. The variance of the ith set of principal component coordinates is equal to λ_i, the ith eigenvalue.
12. The correlation coefficient between the ith and jth sets of principal component coordinates is zero. This follows from the fact that the eigenvectors define a set of orthogonal axes.

7.5 PROBLEMS WITH CONSTRUCTING VARIATION DIAGRAMS

A word of warning must be inserted here concerning the calculation of principal component coordinates for use in constructing variation diagrams, especially when a correlation matrix has been used. Trochimczyk & Chayes (1978) have pointed out, that if the wrong method is used to calculate the sets of "principal component coordinates", e.g. by using equation 7.16 instead of 7.17, correlations develop between them, which conflicts with point 12 above. This is shown in Table 7.4, where correlation matrices are given for two sets

TABLE 7.4

Example of the correlations induced by incorrectly calculating the "principal component coordinates" for the pyroxene data of Table 7.1. Note the strong correlations developed between certain sets of "principal component coordinates" and the change in the relative weighting of the variances of the "principal component coordinates", which are equivalent to the "eigenvalues". Abbreviations:- Di = diopside; Hd = hedenbergite; En = enstatite; Fs = ferrosilite.

	Method A using equation 7.16 with eigenvectors extracted from the correlation matrix					Method B using equation 7.17 with eigenvectors extracted from the variance-covariance matrix				
Anal.	1st	2nd	3rd	4th	5th	1st	2nd	3rd	4th	5th
1	-53.79	-8.31	8.12	25.44	31.76	2.43	2.06	0.89	-1.30	0.40
2	-23.34	-10.62	7.91	37.76	37.93	-0.61	0.97	-0.30	0.20	-0.21
3	-4.37	-16.27	5.04	46.82	43.64	-2.59	0.64	0.75	0.54	-1.19
4	-41.26	13.47	-4.59	31.95	29.69	1.72	-1.17	0.01	-0.69	0.77
5	-14.68	6.30	-7.55	42.88	37.54	-1.04	-1.67	0.97	0.25	-0.52
6	-37.29	11.23	-3.30	31.97	31.15	1.28	-1.06	-0.18	-0.16	0.57
7	-20.21	6.09	-1.70	37.49	35.76	-0.51	-1.22	-0.63	0.93	-0.00
8	-28.73	-0.86	2.91	34.26	35.33	0.16	0.03	-0.29	0.37	0.05
9	-18.83	-0.47	6.34	41.55	35.25	-0.81	-0.34	-2.12	0.33	0.56
10	-35.90	-7.83	4.87	32.46	35.67	0.69	1.13	0.88	-0.44	-0.17
11	-21.69	-8.83	5.61	39.73	37.94	-0.72	0.64	0.01	-0.04	-0.26

Variance of "principal component coordinates" or "eigenvalues"

	1st	2nd	3rd	4th	5th	1st	2nd	3rd	4th	5th
	192.66	94.75	29.87	37.05	14.90	2.06	1.42	0.82	0.39	0.32

"Principal component coordinates" for theoretical end-member pyroxenes

	1st	2nd	3rd	4th	5th	1st	2nd	3rd	4th	5th
Di	-43.47	14.26	-10.75	30.64	30.80	1.97	-1.35	2.07	-0.97	0.13
Hd	-12.88	5.51	-9.71	43.68	38.69	-1.23	-1.73	1.77	0.20	-0.86
En	-59.31	-9.01	7.04	23.98	31.15	2.95	2.34	1.70	-1.85	0.34
Fs	2.01	-19.90	4.75	50.09	45.91	-3.30	0.73	1.22	0.56	-1.58

Correlation matrix for the "principal component coordinates" of Method A

	1st	2nd	3rd	4th	5th
1st	1.0000	-0.2644	-0.0525	0.9820	0.8545
2nd	-0.2644	1.0000	-0.8436	-0.2556	-0.6939
3rd	-0.0525	-0.8436	1.0000	-0.0590	0.3081
4th	0.9820	-0.2556	-0.0590	1.0000	0.8259
5th	0.8545	-0.6939	0.3081	0.8259	1.0000

Correlation matrix for the "principal component coordinates" of Method B

	1st	2nd	3rd	4th	5th
1st	1.0000	0.1373	0.1367	-0.8140	0.7859
2nd	0.1373	1.0000	0.2848	-0.4495	-0.1962
3rd	0.1367	0.2848	1.0000	-0.4129	-0.4731
4th	-0.8140	-0.4495	-0.4129	1.0000	-0.4713
5th	0.7859	-0.1962	-0.4731	-0.4713	1.0000

of "principal component coordinates" which have been calculated incorrectly. As can be seen, some of the correlation coefficients are extremely large. Unfortunately, the use of the incorrect method A as shown in Table 7.4 is fairly common, e.g. Saxena (1969), Saxena & Ekström (1970), Saxena & Walter (1974), Till & Colley (1973). Although method B in Table 7.4 is also a theoretical possibility, it is extremely unlikely that it would ever be used accidentally. These induced correlations were also noted by Butler (1976), who intimated that they could be due to using closed data. However, the reason is simply due to using the wrong equations, as the calculated "principal component coordinates" are then no longer the coordinates of the data projected onto the orthogonal eigenvectors, but are the coordinates of the data projected onto a new set of vectors, which are related to the eigenvectors by an oblique rotation.

It should, therefore, be obvious that if principal component coordinates are to be used to display data, or to construct variation diagrams, it is important to calculate them correctly, otherwise there is no object in using principal components analysis in the first place, as one may just as well choose arbitrary vectors on which to project the data. Admittedly a diagram can always be produced by using the wrong method of calculation, but it will give a misleading impression of the spatial relationship of the data, as shown in Figs. 7.3e and 7.3f. Furthermore, the diagrams will also not display the maximum amount of variance, as can be seen in Table 7.4, where the sum of the variances of the first two sets of "principal component coordinates" has decreased, compared to the sum of the first two eigenvalues of Tables 7.2 and 7.3. Note also that the 5th "eigenvalues" are no longer zero and that the set of "eigenvalues" from method A are no longer in decreasing order of magnitude. Some other examples of diagrams produced incorrectly in this way, can be found in Saxena (1969), Saxena & Ekström (1970), Saxena & Walter (1974) and Till & Colley (1973).

7.6 USE AS A "SURFACE-FITTING" PROCEDURE

Gnanadesikan (1977) has pointed out that principal components analysis can be used as a "surface-fitting" procedure in p-dimensional space, by examining the eigenvector of the smallest eigenvalue. It has already been noted in the preceding section, that if the eigenvectors and eigenvalues are extracted from a variance-covariance matrix computed for closed data with a constant sum, the pth eigenvector defines the normal to the plane of constant composition (equation 7.10), which is the surface in which all the data lies. Similarly, one could test the supposition that the data are confined to a polynomial surface in p-dimensional space, by introducing suitable squared and cross-product terms as variables in the principal components analysis, in the same way that similar terms are introduced into multiple linear regression to perform polynomial regression (section 6.2). If the pth eigenvalue is then found to be small, the pth eigenvector gives the equation of the surface in which most of the data lies. To illustrate the method, Gnanadesikan (1977) gives some examples of artificial 3-dimensional data that are closely confined to a paraboloid and spherical surface. It must be pointed out, however, that like polynomial regression, there is no way in which the type of surface can be deduced, other than by trial and error.

This approach to principal components analysis opens up the possibilty of using the method to detect outliers or anomalies in the data set, by examining the pth set of principal component coordinates for large departures from the mean value, in the same way that the residuals in multiple regression analysis can be used (section 6.2). Again, this could be particularly important in detecting anomalies in geochemical prospecting.

This method of "surface-fitting" was also advocated by Hey (1969) for dealing with multivariate mineralogical data when all the variables are subject to error.

7.7 SIGNIFICANCE TESTS

Although principal components analysis usually makes no assumption about the variables, if it is assumed that the sample has been drawn from a multivariate normally distributed population, with a specific variance-covariance matrix, certain large-sample significance tests are available. Probably the most important of these, from the petrological point of view, is the test for deciding if a particular eigenvector extracted from a variance-covariance matrix is equal to some specified vector. If $\mathbf{D}$ is the large-sample variance-covariance matrix, based on n observations, $\mathbf{w}$ is the specified vector and λ_i is the eigenvalue associated with the eigenvector to be tested, Anderson (1963) has shown that, when the null hypothesis $H_0 : \mathbf{w} = \mathbf{v}_i$ is true:-

$$\chi^2 = (n - 1)\left[\lambda_i \mathbf{w}^* \mathbf{D}^{-1} \mathbf{w} + \frac{1}{\lambda_i} \mathbf{w}^* \mathbf{D} \mathbf{w} - 2\right]$$

(7.19)

where $\mathbf{w}^*$ is the transpose of $\mathbf{w}$, is distributed as chi-square with (p - 1) d.f. Confidence limits may also be assigned to the eigenvalues but, as Marriott (1974) has pointed out, there is no test to decide if some of the principal components can be ignored.

Significance tests for eigenvectors and eigenvalues extracted from the correlation matrix are much more limited due to the exceedingly complicated mathematics. However, Trochimczyk & Chayes (1977) have empirically investigated the terms of eigenvectors derived from correlation matrices and concluded that they are subject to considerable sample variance and that care should be exercised in the interpretation of the values. Unfortunately, they reported no results for eigenvectors extracted from variance-covariance matrices.

CHAPTER 8

FACTOR ANALYSIS

8.1 INTRODUCTION

Factor analysis is a general name for a variety of procedures to examine the correlations within a set of data, and was originally proposed by Spearman (1904) to explain the relationships between certain postulated factors in the field of educational psychology. However, inspite of it being the oldest of the main multivariate techniques, there is still a considerable lack of agreement in the literature over the meaning of the term "factor analysis". This is partly due to the fact that many authors have widened the term to include techniques such as principal components analysis which, unlike factor analysis, assumes no well-defined mathematical model (Marriott, 1974). As a result, one can find examples in the geological literature allegedly dealing with factor analysis when only principal components analysis has been used and *vice versa*. However, as pointed out by many authors (e.g. Francis, 1974; Jöreskog et al., 1976; Marriott, 1971; Temple, 1978), there are distinct differences between the concepts and assumptions of principal components analysis, which is variance-orientated, and factor analysis, which is covariance- or correlation-orientated.

Even the validity of the use of factor analysis is still a controversial topic. For example, Francis (1974) in a critique of the use of factor analysis comments that "Little has been written in statistical journals regarding factor analysis, and few textbooks in statistics treat the subject seriously. Yet we find that it is one of the most widely used 'statistical' procedures. In my experience it is also the most misunderstood and misapplied". He also comments (p.10) that "perhaps the principal controversy regarding factor analysis surrounds the frequent interpretation of estimated factors as being real, as having causal implications regarding the observed phenomena".

Similarly, following closely on the heels of the text of Jöreskog et al., (1976) on the subject of geological factor analysis, Temple (1978) critically examined the use of factor analysis in geology and came to the conclusion that, *as currently practiced*, it should not continue to be used. Unfortunately, among many people factor analysis has become a method "for making invisible influences visible" (Cattell, 1965, p.191) and appears to have acquired the mystique of a religion - you either believe it or you don't. In the hands of the uninitiated, factor analysis can surely claim to be the perfect example of a numerical "black-box" technique, into which data is put at one end and, depending upon which button you push, different answers come out of the other end. As Blackith & Reyment (1971, p.201) state "It is very hard to discuss factor analysis .... without generating more heat than light".

In a comparison of several commonly used computer packages which perform factor analysis, Francis (1974) warns that different results may be obtained from programs that allegedly perform the same task. Using some artificially generated data, in which the number of factors was known, he also showed that some of the methods produced quite spurious, but plausible, results when the assumed number of factors was different from the number known to be actually present. As he comments (p.25) "There is little doubt that an imaginative user could come up with a plausible explanation for the high estimated loadings on these non-existent factors". Anybody contemplating factor analysis would, therefore, by wise to try several methods in order to assure that the answers are reasonably consistent. If they are not, there is no logical reason for chosing one answer in preference to another, apart from ones preconceived ideas. As Miesch (1976d, p.46) comments "It will always be possible to derive a number of models that are mathematically satisfactory. The ultimate test of the model is its geological plausibility". In this respect, factor analysis is similar to the addition-subtraction problem discussed in section 5.4, as both methods provide an infinite number of solutions.

The geological applications of factor analysis are generally confined to two methods; one is called *R-mode factor analysis* and investigates the relationships between variables, while the other is called *Q-mode factor analysis* and investigates the relationships between the objects. While some authors (e.g. Blackith & Reyment 1971; Francis, 1974; Marriott, 1974) have questioned whether R-mode factor analysis has ever achieved anything that could not have been done by principal components analysis, an extension of the Q-mode method has proved to be a useful technique in petrology.

8.2 R-MODE FACTOR ANALYSIS

The basic aim of R-mode factor analysis is to explain the covariances or correlations between the *observed* variables in terms of a linear combination of a specified number of *unobservable* variables, called *factors*.

Expressed in terms of the observed variables, which are usually assumed or scaled to have zero means, the factor model may be written as:-

$$x_{ij} = \sum_{k=1}^{m} a_{jk}f_{ik} + e_{ij} \qquad (8.1)$$

where x_{ij} is the jth variable of the ith object, m is the specified number of factors thought to be responsible for the observed correlations, a_{jk} is the kth *factor loading* for the jth variable, f_{ik} is the kth *factor score* of the ith object, and e_{ij} is a residual error term. This means that each variable is regarded as a linear sum of what are known as *common factors*, plus an error term, which is known as a *unique factor*.

Unfortunately, there is no unique set of factor scores and loadings that will satisfy equation 8.1, as it is indeterminate. For example, if the scale of one of the factor scores is changed with a corresponding change in the factor loading, then the result will not be changed, so that some type of standardization is required. Secondly, if the number of factors, m, is greater than 1, there is another type of indeterminacy. We can express equation 8.1 in matrix form as:-

$$\mathbf{x} = \mathbf{A}\mathbf{f} + \mathbf{e} \tag{8.2}$$

where **x** is a px1 vector of variables for one object, **f** is a mx1 vector of the factor scores of the object, **A** is the pxm matrix of factor loadings, which is also referred to as the *factor pattern*, and p is the number of variables. If we then replace **f** by the linear combination **Bf**, where **B** is a mxm transformation matrix, we obtain:-

$$\mathbf{x} = \mathbf{A}\mathbf{B}\mathbf{f} + \mathbf{e} \tag{8.3}$$

which will leave the result unchanged, as long as $\mathbf{A}\mathbf{B}\mathbf{f} = \mathbf{A}\mathbf{f}$. However, as there are an infinite number of matrices **B** that will satisfy this condition, it follows that there are also an infinite number of sets of factor loadings and scores that will satisfy the factor model (Marriott, 1974). It is this indeterminacy that allows the factors to be rotated without effecting their power to estimate the original data.

The factor model of equation 8.1 may be also be expressed in terms of **D**, which may be either a pxp variance-covariance matrix as defined by equation 7.1, or a correlation matrix as defined by equation 7.2, as:-

$$\mathbf{D} = \mathbf{A}\Phi\mathbf{A}^* + \Psi \tag{8.4}$$

where $\mathbf{A}^*$ is the transpose of **A**, Φ (Greek letter capital phi) is the mxm covariance matrix of the factors, and Ψ (Greek letter capital psi) is the pxp residual covariance matrix. For this model, the covariance between the ith variable and the jth factor is given by the corresponding element of the pxm matrix product $\mathbf{A}\Phi$, which is known as the *factor structure*.

However, if the factors are uncorrelated and orthogonal, Φ becomes an identity matrix, so that equation 8.4 simplifies to:-

$$\mathbf{D} = \mathbf{A}\mathbf{A}^* + \Psi \tag{8.5}$$

The covariance between the ith variable and the jth factor is then given by a_{ij}, the element of the factor loading matrix. In this case the factor pattern and the factor structure are the same.

The assumptions that have been made in this model so far are that:- i) the data can be explained in terms of a small number of factors; ii) the variables are linear combinations of common factors plus a unique factor; iii) the residual error terms are uncorrelated with both the factors and each other;

iv) the factors generate all the correlations in the data, but only part of the variances, which are called *communalities*; and   v) the remaining part of the variances are generated by the diagonal terms of Ψ or the *uniquenesses*.

The details of the solution of these equations to provide estimates of $\mathbf{A}$ and Ψ, is beyond the scope of this book but further information may be found in many texts, e.g. Jöreskog et al. (1976), Lawley & Maxwell (1971), Morrison (1967). Suffice it to say that several methods are available, all of which are iterative. Some of the methods involve the additional assumption that the data is multivariate normally distributed and provide significance tests for assessing the fit of the model. Similarly, the determination of the factor scores is a complicated procedure and again several methods are available for their estimation.

Once factors loadings have been determined, $h_j{}^2$, the *communality* of the jth variable can be calculated as:-

$$h_j{}^2 = \sum_{k=1}^{m} a_{jk}{}^2 \qquad (8.6)$$

The *uniqueness* of the jth variable, c_{jj}, which is a diagonal term of matrix Ψ, is then given by:-

$$c_{jj} = d_{jj} - h_j{}^2 \qquad (8.7)$$

where d_{jj} is the variance of the jth variable and is a diagonal term of matrix $\mathbf{D}$. If the data have been standardized, d_{jj} is, of course, 1.

One of the problems with factor analysis is that there is no generally agreed way to decide how many factors to use. From the practical point of view it is probably best to limit the number of factors to 2 or 3, as this number is capable of graphical representation (see section 8.4.1). If considerably more than 3 factors are required to obtain reasonable communalities, their interpretation is often difficult, which may suggest that the factor model is not an appropriate model to use.

What has been described so far is what is often referred to as *"true" factor analysis*, as it involves uniquenesses that are assumed to have definite values. However, "true" factor analysis has rarely been used in geology.

8.2.1 Simplfied version commonly used in geology

In geological factor analysis the uniquenesses are usually assumed to be zero, so that equation 8.5 reduces to:-

$$\mathbf{D} = \mathbf{A}\mathbf{A}^{*} \qquad (8.8)$$

This simplified version of factor analysis is often referred to as *principal factor analysis*. If the number of factors, m, is assumed to be equal to the number of variables, p, one of the solutions to equation 8.8 is given by:-

$$\mathbf{A} = \mathbf{V}\Lambda^{\frac{1}{2}} \qquad (8.9)$$

where **V** is the matrix of eigenvectors of **D**, and **Λ** is an mxm matrix containing the eigenvalues of **D** in its diagonal terms, for if we substitute **A** in equation 8.8 we obtain:-

$$\mathbf{D} = \mathbf{V\Lambda}^{\frac{1}{2}}(\mathbf{V\Lambda}^{\frac{1}{2}})^* = \mathbf{V\Lambda}^{\frac{1}{2}}\mathbf{\Lambda}^{\frac{1}{2}}\mathbf{V}^* = \mathbf{V\Lambda V}^* \tag{8.10}$$

which is one of the definitions of the eigenstructure of a matrix (equation 7.6). It is for this reason that principal components analysis is regarded by some authors as part of factor analysis, but it is such a special case that it is "more appropriate, both historically and conceptually, to consider principal components analysis as a seperate technique" (Temple, 1978, p.383). The estimated factor loadings given by equation 8.9 are often referred to as *initial* or *unrotated factor loadings*.

When the uniquenesses are assumed to be zero, the estimation of the factor scores, unlike that in "true" factor analysis, is relatively simple and similar to the calculation of principal component coordinates. If the factor loadings have been calculated using equation 8.9, the factor scores are given by:-

$$\hat{\mathbf{F}} = \mathbf{YAV}^{-1} \tag{8.11}$$

where $\hat{\mathbf{F}}$ is the nxm matrix of estimated factor scores, $\mathbf{\Lambda}^{-1}$ is the inverse of **Λ**, and **Y** is the nxp matrix of adjusted variables, the terms of which are defined as follows. If unstandardized data are used and **D** is a variance-covariance matrix then the terms of the matrix **Y** are given by y_{ij} where:-

$$y_{ij} = x_{ij} - \bar{x}_j \tag{8.12}$$

and $\bar{x}_j$ is the mean of the jth variable. If, however, the variables are standardized and **D** is a correlation matrix then:-

$$y_{ij} = (x_{ij} - \bar{x}_j) / SDx_j \tag{8.13}$$

where SDx_j is the standard deviation of the jth variable. When the original data is rewritten in the form of equation 8.12 or 8.13, it also follows that **D** is given by:-

$$\mathbf{D} = \mathbf{Y}^*\mathbf{Y} / (n - 1) \tag{8.14}$$

where $\mathbf{Y}^*$ is the transpose of **Y**. The use of this expression for calculating a variance-covariance or correlation matrix is, however, only practical if all the data can be held in the computer core at one time.

If **A** is eliminated from equation 8.11 we obtain:-

$$\hat{\mathbf{F}} = \mathbf{YV\Lambda}^{\frac{1}{2}}\mathbf{\Lambda}^{-1} = \mathbf{YV\Lambda}^{-\frac{1}{2}} \tag{8.15}$$

The relationship between these factor scores and principal component coordinates now becomes clear, as equations 7.15 and 7.17 can be written in matrix form as:-

$$\mathbf{Z} = \mathbf{Y}\,\mathbf{Y} \tag{8.16}$$

where $\mathbf{Z}$ is the nxm matrix of principal component coordinates which, by substitution in equation 8.15, gives the relationship that:-

$$\hat{\mathbf{F}} = \mathbf{Z}\boldsymbol{\Lambda}^{-\frac{1}{2}}$$

$$(8.17)$$

In other words the factor scores are the principal component coordinates divided by the square roots of their respective eigenvalues. This means that the factor scores are standardized to unit variance as, by definition, the variance of a set of principal component coordinates is equal to the corresponding eigenvalue (see section 7.4). It follows, therefore, that if the factor scores are used to plot the location of objects in p-dimensional space, their relative locations will be altered due to the standardization of the factor scores. If a correlation matrix has been used, as is usually the case, this change in the relative locations of the objects will be in addition to that introduced when the original data were standardized, as discussed in section 7.3.

Although equations 8.9 and 8.11 are only one way of estimating the initial factor loadings and scores, they are commonly used in the geological literature. Another solution (Jöreskog et al., 1976) is that given by the equations:-

$$\mathbf{A} = \mathbf{V}$$

$$(8.18)$$

and

$$\hat{\mathbf{F}} = \mathbf{Y}\mathbf{V}$$

$$(8.19)$$

which is, of course, principal components analysis *sensu stricto*, in which the eigenvectors are called factor loadings and the principal component coordinates are called factor scores.

Numerical example

The initial factor loadings for a 2-factor model, determined using equation 8.9, are given in Table 8.1 for the pyroxene data of Table 7.1. As can be seen, the initial factor loadings are simply the eigenvectors of Tables 7.2 and 7.3 multiplied by the square root of their respective eigenvalue. The initial factor scores are calculated using equation 8.11. Multiplying these by the square roots of their respective eigenvalues (equation 8.17) then gives the principal component coordinates as defined by equations 7.15 and 7.17. These should be compared with the results given in Tables 7.2 and 7.3. Note, however, that as the principal component coordinates of Table 7.2 were calculated using equation 7.16, they differ from those of Table 8.1 by a term which is constant for each set of coordinates.

The two sets of factor scores in Table 8.1 are plotted against each other in Fig. 8.1, together with the pyroxene quadrilateral, for comparison with the principal component coordinate plots of Fig. 7.3c and 7.3d. When compared with the principal component coordinate plots, the way in which the 1st set of factor scores are compressed with respect to the 2nd, is noticeable.

TABLE 8.1

Results of an initial or unrotated R-mode factor analysis of the pyroxene data of Table 7.1, using both the variance-covariance and correlation matrices. The intital factor loadings and scores are calculated using equations 8.9 and 8.11, respectively. The principal component coordinates are calculated using equation 8.17.

BASED ON THE VARIANCE-COVARIANCE MATRIX

	Initial Factor loadings			Initial Factor scores		Principal comp. coords.	
	1st	2nd		1st	2nd	1st	2nd
SiO_2	3.052	0.866	1	1.60	1.57	25.47	16.55
R_2O_3	0.097	-0.835	2	-0.45	0.89	-7.16	9.31
FeO	-13.180	2.509	3	-1.86	0.66	-29.48	6.96
MgO	8.082	5.782	4	1.23	-1.12	19.49	-11.78
CaO	1.949	-8.323	5	-0.69	-1.42	-10.94	-14.97
			6	0.95	-0.97	15.07	-10.24
Eigenvalues	252.16	110.46	7	-0.26	-0.95	-4.13	-10.03
			8	0.14	0.05	2.30	0.51
			9	-0.56	-0.16	-8.89	-1.71
			10	0.44	0.89	6.97	9.34
			11	-0.55	0.58	-8.69	6.06
		Variance		1.00	1.00	252.16	110.46

BASED ON THE CORRELATION MATRIX

	Initial Factor loadings			Initial Factor scores		Principal comp. coords.	
	1st	2nd		1st	2nd	1st	2nd
SiO_2	-0.981	0.008	1	-2.03	-1.09	-3.30	-1.36
R_2O_3	0.141	0.697	2	0.26	-0.71	0.42	-0.89
FeO	0.880	-0.464	3	1.48	-1.47	2.40	-1.84
MgO	-0.934	-0.293	4	-0.98	1.28	-1.58	1.60
CaO	0.050	0.882	5	0.87	0.45	1.41	0.56
			6	-0.60	1.16	-0.97	1.46
Eigenvalues	2.63	1.57	7	0.72	0.95	1.18	1.19
			8	0.01	0.15	0.02	0.18
			9	0.66	0.86	1.07	1.08
			10	-0.70	-0.93	-1.13	-1.17
			11	0.30	-0.65	0.49	-0.81
		Variance		1.00	1.00	2.63	1.57

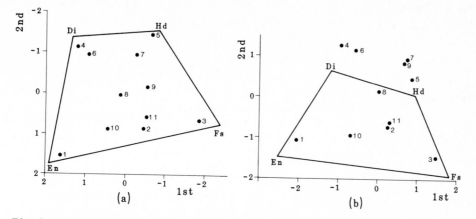

Fig. 8.1. Plots of the 1st versus the 2nd sets of R-mode factor scores of Table 8.1, together with the pyroxene quadrilateral using results calculated from (a) the variance-covariance matrix and (b) the correlation matrix. Note how the plots are stretched along the direction of the 2nd factor scores with respect to the direction of the 1st factor scores, when compared with Fig. 7.3c and 7.3d, respectively.

8.2.2 *Rotation of factors*

After the initial factor loadings have been determined they are usually rotated in some way, in order to get what are often referred to as more "meaningful" factor loadings. To quote Cattell (1965, p.207) "What the experimenter is looking for is a rotation that will cause factors to correspond to certain scientific entities, which can be regarded as explanatory of the observed covariation of the variables". This is usually achieved by applying the concept of *simple structure* (Thurstone, 1947) to the factor loadings i.e. the rotation should provide factor loadings whose absolute magnitudes are either as small or as large as possible, so that they may be more easily interpreted.

Mathematically, each rotated solution is equivalent to selecting a different matrix, **B**, in equation 8.3. Although a large number of maximizing and minimizing criteria have been proposed for achieving simple structure, only two appear to be widely used in factor analysis programs (Jöreskog et al., 1976). One is a rigid orthogonal rotation, called the *varimax procedure*, in which the variance of the squares of the factor loadings is maximized to provide loadings whose absolute magnitudes are either very large or small. Squares are taken to avoid complications with the signs of the loadings. The other is an oblique rotation, called the *promax method*, which starts with the varimax rotation and then minimizes a function which produces factors that tend to be associated with clusters of variables.

However, as Temple (1978, p.382) points out "It is in the rotation of factors that modern factor analysis becomes more demonstrably subjective. Factors can be rotated to any orientation relative to the original variates, and some positions will be more "meaningful" (i.e. more easily interpretable in terms of the variates) than others, but there is no reason to regard such easily interpretable positions as having any special scientific significance".

Although factor analysis can be used to represent the spatial distribution of the data, it is a waste of effort to consider any rotation of the factors if this is the *sole purpose* for performing factor analysis, for as Temple (1978, p.384) again points out "Rotated factors are .... no better than unrotated ones for depicting the spatial relations of individuals or variates. In fact the simplest axes on which to plot both individuals and variates are provided by the eigenvectors of the correlation or covariance matrix".

8.3   Q-MODE FACTOR ANALYSIS

As Q-mode factor analysis is concerned with the relationships between objects, the basic factor model can be written as:-

$$x_{ij} = \sum_{k=1}^{m} a_{ik}f_{jk} + e_{ij} \qquad (8.20)$$

where x_{ij} is the jth variable of the ith object, m the number of factors, a_{ik} is the kth factor loading for the ith *object*, f_{jk} is the kth factor score for the jth *variable*, and e_{ij} is the error term. The only difference between this and equation 8.1 is the interchanged roles of the variables and objects in the factor loadings and scores. By analogy with equation 8.5 the model can also be written in matrix notation as:-

$$\mathbf{D} = \mathbf{A}\mathbf{A}^* + \mathbf{\Psi} \qquad (8.21)$$

where $\mathbf{D}$ is now an nxn variance-covariance matrix, which can be calculated as:-

$$\mathbf{D} = \mathbf{Y}\mathbf{Y}^* / (n - 1) \qquad (8.22)$$

where matrix $\mathbf{Y}$ is defined as for equation 8.11. However, in Q-mode factor analysis, several other similarity measures are also used.

Again in geological applications, the model is usually simplified by omitting the error term so that:-

$$x_{ij} = \sum_{k=1}^{m} a_{ik}f_{jk} \qquad (8.23)$$

or its alternative matrix form:-

$$\mathbf{D} = \mathbf{A}\mathbf{A}^* \qquad (8.24)$$

is the model that is usually used (Imbrie, 1963; Jöreskog et al., 1976; Miesch, 1976b, 1976d).

If the data are compositions, the factor loadings, a_{ik}, can be thought of as relative amounts of end-members that are present in each object, the end-member compositions being defined by the factor scores. In other words, the method can be thought of as a type of multi-object mixing model, where each object is fitted to the same set of end-members. Jöreskog et al. (1976) summarize the objectives of this type of Q-mode factor analysis as follows:-

1. to determine m, the minimum number of end-members (factors) required to adequately explain the data. Like R-mode factor analysis, this is largely subjective.
2. to determine the compositions of the end-members in terms of the p variables. Again due to the indeterminacy in factor analysis, there are an infinite number of possible end-members. In practice, either the most divergent objects are chosen as end-members or theoretical compositions can be used.
3. to describe each object in terms of amounts of the end-members. These weightings are, of course, the factor loadings.

As most compositions are expressed as proportions, Imbrie & Purdy (1962) suggest that covariances and correlation coefficients are not the best similarity measures to use. Instead they suggest that *coefficients of proportional similarity* between objects should be used. These are defined as:-

$$\cos \theta_{ij} = \sum_{k=1}^{p} x_{ik} x_{jk} \Big/ \left[\sum_{k=1}^{p} x_{ik}^2 \sum_{k=1}^{p} x_{jk}^2 \right]^{1/2} \tag{8.25}$$

where i and j are the two objects in question and x_{ik} and x_{jk} are their kth variables. As this measure is the cosine of the angle subtended by the ith and jth object at the origin of measurements of the variables (see equations 5.7 and 5.10), it is also known as the *cosine theta* measure. If the variables for each object are expressed in direction cosine form (section 5.3), by standardizing them so that their sum of squares is 1 by using:-

$$y_{ij} = x_{ij} \Big/ \sqrt{\sum_{j=1}^{p} x_{ij}^2} \tag{8.26}$$

then **D**, the nxn matrix of cosine theta measures, may be computed directly from the matrix expression:-

$$\mathbf{D} = \mathbf{Y} \mathbf{Y}^* \tag{8.27}$$

where **Y** is an nxp matrix whose terms are y_{ij}.

The procedure for the estimation of the factor loadings and scores is then basically the same as that used in R-mode factor analysis. In practice, however, many Q-mode factor analysis programs save computing time by using the generally smaller pxp matrix $\mathbf{Y}^*\mathbf{Y}$ as its positive eigenvalues are identical to those of the larger nxn matrix $\mathbf{Y}\mathbf{Y}^*$ of equation 8.27. The number of non-zero eigenvalues of these matrices cannot be greater than p, the number of variables, assuming that $p < n$. The two sets of eigenvectors are also easily derived from each other (Jöreskog et al., 1976, p.48) as:-

$$\mathbf{V}_q = \mathbf{Y}\,\mathbf{V}_r\,\mathbf{\Lambda}^{-1} \tag{8.28}$$

where $\mathbf{V}_q$ is the nxp matrix of eigenvectors of $\mathbf{YY}^*$, $\mathbf{V}_r$ is the pxp matrix of eigenvectors of $\mathbf{Y}^*\mathbf{Y}$ and $\mathbf{\Lambda}^{-1}$ is a pxp diagonal matrix whose terms are the inverse reciprocals of the eigenvalues.

Numerical example

   The initial stages of Q-mode factor analysis of the pyroxene data of Table 7.1 are given in Table 8.2. Firstly, equation 8.26 is used to form $\mathbf{Y}$, the normalized data matrix, which is then post-multiplied by its transpose, $\mathbf{Y}^*$, to form $\mathbf{D}$ the cosine theta matrix. For example:-

$$\cos \theta_{12} = 0.844 \text{x} 0.835 + 0.027 \text{x} 0.070 + 0.053 \text{x} 0.478 + 0.533 \text{x} 0.263 + 0.003 \text{x} 0.024$$
$$= 0.872$$

Only the first 5 eigenvalues and eigenvectors are given as the remaining eigenvectors are, within rounding-off errors, zero.

8.4 EXTENDED Q-MODE FACTOR ANALYSIS

   For compositional data, Miesch (1976b, 1981) points out that the factor loadings and scores estimated from the nxn matrix $\mathbf{D}$ of cosine theta measures are in an inconvenient form as:- i) the factor loadings for one object will not sum to unity, so that they cannot be considered as proportions and ii) the factor scores are dimensionless quantities and cannot, therefore, be thought of as compositions. In order to overcome these deficiencies, he proposed an important modification of the method which has become known as *extended Q-mode factor analysis*. This scales the factor loadings and scores so that they sum to unity and represent end-member compositions, respectively. Although he concluded that the method could only be used with data in which the variables have a constant sum, Klovan (1981) has recently shown that the procedure can be generalized to data with variable row sums.

   Miesch (1976b) advocates the use of the cosine theta measure (equation 8.25) and makes provision for the data to be scaled or transformed, prior to the calculation of the similarity measure, by means of the general expression:-

$$z_{ij} = (x_{ij} - xmin_j) / (xmax_j - xmin_j) \tag{8.29}$$

where $xmax_j$ and $xmin_j$ are the maximum and minimum values of the jth variable, respectively, if the data are to be transformed to proportions of the total range. If $xmin_j$ is set to zero, the transformation is to proportions of the maximum value, and if $xmax_j$ is then set to 1, the data are not transformed. The transformations allow equal weights to be given to all the variables, in cases where the raw variables have widely different variances, such as with major and minor elements. The implications and effects of some of these transformations are discussed further by Miesch (1980).

   Scale factors, s_k, can then be calculated to convert the unscaled factor

TABLE 8.2

Example of the initial stages of Q-mode factor analysis of the pyroxene data of Table 7.1, giving Y^*, the transpose of the normalized data matrix, and the cosine theta matrix, D, which is calculated as YY^*. The first 5 eigenvectors and eigenvalues of D are also given.

Transpose of normalized data matrix, Y^*

	1	2	3	4	5	6	7	8	9	10	11
SiO_2	0.844	0.835	0.700	0.891	0.817	0.891	0.850	0.878	0.858	0.872	0.842
R_2O_3	0.027	0.070	0.020	0.076	0.032	0.085	0.101	0.082	0.152	0.028	0.059
FeO	0.053	0.478	0.711	0.026	0.450	0.096	0.388	0.331	0.454	0.288	0.487
MgO	0.533	0.263	0.053	0.280	0.018	0.278	0.133	0.286	0.121	0.388	0.215
CaO	0.003	0.024	0.022	0.348	0.360	0.336	0.314	0.175	0.140	0.067	0.064

Cosine theta matrix $D = YY^*$

	1	2	3	4	5	6	7	8	9	10	11
1	1.000	0.872	0.658	0.906	0.725	0.908	0.812	0.914	0.817	0.959	0.853
2	0.872	1.000	0.941	0.843	0.912	0.877	0.945	0.976	0.979	0.972	0.998
3	0.658	0.941	1.000	0.666	0.901	0.716	0.887	0.871	0.937	0.838	0.950
4	0.906	0.843	0.666	1.000	0.872	0.997	0.922	0.938	0.871	0.919	0.850
5	0.725	0.912	0.901	0.872	1.000	0.899	0.988	0.937	0.962	0.874	0.936
6	0.908	0.877	0.716	0.997	0.899	1.000	0.946	0.959	0.902	0.938	0.883
7	0.812	0.945	0.887	0.922	0.988	0.946	1.000	0.976	0.981	0.929	0.960
8	0.914	0.976	0.871	0.938	0.937	0.959	0.976	1.000	0.975	0.987	0.978
9	0.817	0.979	0.937	0.871	0.962	0.902	0.981	0.975	1.000	0.940	0.988
10	0.959	0.972	0.838	0.919	0.874	0.938	0.929	0.987	0.940	1.000	0.964
11	0.853	0.998	0.950	0.850	0.936	0.883	0.960	0.978	0.988	0.964	1.000

Eigenvectors

	1st	2nd	3rd	4th	5th
1	0.282	-0.465	0.486	-0.123	0.123
2	0.309	0.157	0.295	0.113	-0.132
3	0.280	0.569	0.169	-0.274	0.127
4	0.293	-0.417	-0.324	0.020	0.372
5	0.299	0.217	-0.476	-0.452	0.282
6	0.300	-0.337	-0.289	0.019	-0.071
7	0.309	0.080	-0.320	0.056	-0.621
8	0.314	-0.055	0.037	0.035	-0.466
9	0.310	0.191	-0.051	0.787	0.279
10	0.309	-0.142	0.305	-0.259	-0.048
11	0.310	0.187	0.187	0.027	0.211

Eigenvalues

1st	2nd	3rd	4th	5th
10.088	0.613	0.282	0.015	0.002

As accumulative percentages

91.7	97.3	99.8	100.0	100.0

loadings (a_{ik}) and scores (f_{jk}) into *compositional loadings* (a_{ik}') and *compositional scores* (f_{jk}'), respectively, by using the expression:-

$$s_k = (c - \sum_{j=1}^{p} xmin_j) / \sum_{j=1}^{p} f_{jk} (xmax_j - xmin_j) \qquad (8.30)$$

where c is the constant sum of the variables, usually 100. The compositional loadings or proportions of the end-members are then given by:-

$$a_{ik}' = (a_{ik} / s_k) / \sum_{k=1}^{m} (a_{ik} / s_k) \qquad (8.31)$$

and the compositional scores or end-member compositions by:-

$$f_{jk}' = s_k f_{jk} (xmax_j - xmin_j) + xmin_j \qquad (8.32)$$

Note that this nomenclature differs slightly from that of Miesch (1976b, 1976d).

Numerical example

The initial unrotated factor loadings and scores are given in Table 8.3, for a 2-factor model, using the data of Table 8.2 and equations 8.9 and 8.11. Note how the factor loadings do not sum to zero, so that they cannot be considered as proportions, and that the factor scores are not compositions. However, when the factor loadings and scores are scaled using equations 8.30, 8.31 and 8.32, the composition loadings then sum to unity and the composition scores then represent compositions. The 1st composition scores are, in fact, very close to the mean pyroxene composition given in Table 7.1. The 2nd composition scores are, however, way outside any mineralogical composition space, as some of the values are negative and some are considerably greater than 100.

8.4.1 Graphical representation and rotation of factors

A geometric consequence of using the cosine theta measure is that each object is represented by a vector of unit length in the p-dimensional variable space. The 1st initial or unrotated factor then approximates the mean position of the cluster of unit vectors in space. This can be seen in Table 8.3, where the loadings on the 1st factor are all large and positive. The 2nd unrotated factor then defines the direction of maximum "spread" of the vectors at right angles to the 1st and so on, in a manner analogous to that in principal components analysis. For a 2-factor model each original vector of unit length can then be projected as a new vector into the factor space, by plotting the factor loadings against each other to produce a diagram, as shown in Fig. 8.2 for the pyroxene data of Table 7.1. The lengths of the projected vectors are then the square roots of the communalities, as defined by the Q-mode equivalent of equation 8.6, which is:-

TABLE 8.3

Initial factor loadings and scores for a 2-factor model of the pyroxene data from Table 8.2, together with the scale factors that were used to calculate the composition loadings and scores, which respresent proportions and compositions, respectively. Note how the 1st composition score is similar to the mean pyroxene composition from Table 7.1.

	Factor						Composition					
	loadings			scores			loadings			scores		
	1st	2nd		1st	2nd		1st	2nd		1st	2nd	
1	0.895	-0.364	SiO_2	0.881	-0.178	1	1.005	-0.005	SiO_2	50.96	-845.12
2	0.980	0.123	R_2O_3	0.070	-0.005	2	0.998	0.002	R_2O_3	4.06	-25.00
3	0.890	0.446	FeO	0.358	0.842	3	0.994	0.006	FeO	20.71	4009.69
4	0.929	-0.326	MgO	0.243	-0.485	4	1.004	-0.004	MgO	14.05	-2306.79
5	0.951	0.170	CaO	0.177	-0.154	5	0.998	0.002	CaO	10.21	-732.78
6	0.952	-0.264				6	1.003	-0.003			
7	0.983	0.063				7	0.999	0.001			
8	0.999	-0.043	Scale factors			8	1.001	-0.001			
9	0.984	0.149		57.83	4759.28	9	0.998	0.002			
10	0.980	-0.111				10	1.001	-0.001			
11	0.984	0.146				11	0.998	0.002			

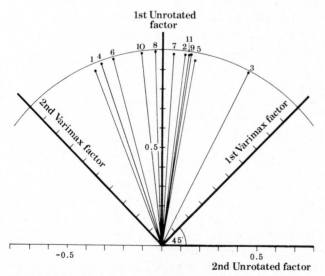

Fig. 8.2. Projection of the object vectors of the pyroxenes of Table 7.1 into the space of the two unrotated factors of Table 8.3, constructed by plotting the Q-mode factor loadings against each other. Note how the object vectors cluster around the 1st unrotated factor. With the varimax rotation of two factors, the angle of rotation is 45, so that in this example, the object vectors are contained completely within the two varimax factors. Comparison of the relative location of the pyroxenes in this diagram and Figs. 7.3 and 8.1, indicate that this 2-dimensional diagram is essentially a projection of the data onto the 1st R-mode factor.

$$h_j{}^2 = \sum_{k=1}^{m} a_{jk}{}^2 \tag{8.33}$$

If a 3-factor model is used, the data can be represented either as vectors in three dimensions, or on a stereographic projection (Miesch, 1976c, 1976d), or as the composition loadings sum to unity, on a triangular diagram as shown in Fig. 8.3.

However, if the factor loadings are thought of as the amounts of the end-members present in each object, it will always be necessary to rotate the factors so that they enclose the vectors. The first rotation usually tried is a rigid varimax rotation which, in the case of a 2-factor model, simply rotates the factors through 45°, as illustrated in Fig. 8.2. With more factors, however, the angle of rotation is determined entirely by the factor loadings. If a rigid rotation does not produce satisfactory results, then an oblique rotation may be tried. For this Miesch (1976c, 1976d) suggests a simple method, originally due to Imbrie (1963) and also described by Jöreskog et al., (1976). An mxm matrix $\mathbf{T}$ is formed, in which each row is an end-member either chosen theoretically or selected as an extreme vector from the varimax factor loadings by choosing rows that have the highest loadings in each column. The oblique factor loadings are then calculated as:-

$$\mathbf{F_O} = \mathbf{F_V}\mathbf{T}^{-1} \tag{8.34}$$

where $\mathbf{F_O}$ is the nxm matrix of oblique factor loadings, $\mathbf{F_V}$ is the nxm matrix of varimax loadings and $\mathbf{T}^{-1}$ is the inverse of $\mathbf{T}$. If the initial choice of $\mathbf{T}$ does

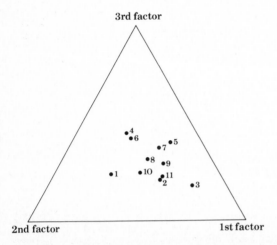

Fig. 8.3. A triangular plot of the Q-mode varimax composition loadings, which sum to 1 for each object, using the 3-factor model of the pyroxene data from Table 8.5. The objects may equally well be plotted on a stereogram, but a triangular diagram is easier to construct. Note how the relative locations of the pyroxenes is now similar to Figs. 7.3 and 8.1.

not give satisfactory results, Miesch (1976c) suggests some iterative methods to find other solutions. More recently, however, Full et al. (1981) have outlined a new iterative method for formalizing the search for "realistic" end-members, which attempts to locate a polytope in p-dimensional space whose vertices are the end-members and whose sides contain at least one data point. The method not only provides positive composition scores within pre-defined limits but also positive composition loadings.

Numerical example

The results of a rigid varimax rotation on the data of Table 8.3 are given in Table 8.4. Note how the factor loadings are now all positive, as the object vectors are all contained within the two factors as shown in Fig. 8.2. The two end-member composition scores are now "extreme" pyroxene compositions but are unsatisfactory due to their negative values. This suggests that an oblique rotation may provide better results. Note that the two factors are still orthogonal, as the cross product sum of their factor scores is zero.

Similarly, the composition loadings and scores derived from a rigid varimax rotation for a 3-factor model are given in Table 8.5. As can be seen the loadings are completely different to those of the 2-factor model in Table 8.4. The composition scores, however, are again mineralogically unsatisfactory due to the negative values. The 1st and 2nd sets of loadings and scores for the *unrotated* 3-factor model are, of course, identical to those of the 2-factor model. For further examples of oblique rotations the reader can refer to Miesch (1976d).

TABLE 8.4

Varimax rotated factor loadings for a 2-factor model, using the initial factor loadings of Table 8.3. Note how the composition scores are more realistic than those of Table 8.3, but the negative values suggest an oblique rotation may give better results.

	F a c t o r						C o m p o s i t i o n					
	loadings		scores				loadings		scores			
	1st	2nd		1st	2nd		1st	2nd		1st	2nd	
1	0.375	0.890	SiO$_2$	0.498	0.749	1	0.302	0.698	SiO$_2$	40.20	61.98	
2	0.780	0.606	R$_2$O$_3$	0.046	0.053	2	0.569	0.431	R$_2$O$_3$	3.71	4.42	
3	0.944	0.314	FeO	0.849	-0.342	3	0.755	0.245	FeO	68.60	-28.35	
4	0.427	0.888	MgO	-0.171	0.514	4	0.330	0.670	MgO	-13.81	42.60	
5	0.792	0.552	CaO	0.016	0.234	5	0.595	0.405	CaO	1.29	19.35	
6	0.487	0.860				6	0.367	0.633				
7	0.740	0.651				7	0.538	0.462				
8	0.676	0.737	Scale factors			8	0.485	0.515				
9	0.801	0.590		80.80	82.79	9	0.582	0.418				
10	0.615	0.771				10	0.449	0.551				
11	0.799	0.592				11	0.580	0.420				

TABLE 8.5

Varimax rotated composition loadings for a 3-factor model, calculated from the data of Table 8.3. Note how the values of the 1st and 2nd loading scores are different from those of Table 8.4. The data for this model is plotted in Fig. 8.3.

	Composition loadings				Composition scores		
	1st	2nd	3rd		1st	2nd	3rd
1	0.263	0.501	0.236	SiO_2	39.59	70.61	48.36
2	0.499	0.289	0.212	R_2O_3	2.55	-0.13	9.87
3	0.664	0.156	0.180	FeO	80.50	-28.06	-14.94
4	0.222	0.330	0.448	MgO	-10.31	86.51	-18.99
5	0.450	0.150	0.400	CaO	-12.33	-28.93	75.70
6	0.259	0.315	0.426				
7	0.413	0.209	0.377				
8	0.395	0.298	0.306	Scale factors			
9	0.478	0.225	0.297		91.17	118.72	93.54
10	0.386	0.366	0.248				
11	0.499	0.264	0.237				

8.4.2 *Factor-variance diagrams*

To determine the number of factors that should be used in the factor model, Miesch (1976d) advocates the construction of a factor-variance diagram. This is done for each variable, by calculating the square of the correlation coefficient, called the *coefficient of determination*, between the actual values of the variables, x_{ij}, and the values of the variables estimated from the factor model, $\hat{x}_{ij}$, by using the expression:-

$$\hat{x}_{ij} = \sum_{k=1}^{m} a_{ik}' f_{jk}' \qquad (8.35)$$

The values of the coefficient of determination for the jth variable, r_j^2, may be calculated either by using an equation such as 4.4 or by using the approximation that:-

$$r_j^2 = \left[\text{var } x_j - \text{var } (x - \bar{x})_j \right] / \text{var } x_j \qquad (8.36)$$

where var x_j is the variance of the jth observed variable and var $(x - \bar{x})_j$ is the variance of the differences between x_{ij} and $\hat{x}_{ij}$ for the jth variable.

The values of r_j^2 are independent of any rotations that may have taken place, so that they may be calculated easily from the unrotated factors. The factor-variance diagram is then constructed by plotting the values of r_j^2 for each variable against the number of factors that could be used in the model.

Numerical example

The values of r_j^2 calculated for the pyroxene data of Table 7.1 are given in Table 8.6. When the number of factors is 1, all the values of r_j^2 are zero. When the number of factors is equal to the number of variables, then all the r_j^2 are 1, as the factor model will then reproduce the original data exactly. These values are plotted in a factor-variance diagram in Fig. 8.4. The number of factors to be used in the model can then be chosen as the minimum number that explains most of the variance in the variables. In this example, a 3-factor model would probably be chosen as, except for R_2O_3, this would account for most of the variance in the variables.

An instructive example of the use of a much more complicated factor-variance diagram in an actual research situation is given by Stuckless & Miesch (1981). Using 38 major and minor oxides (including rare earths) determined on 29 granite specimens from Wyoming, they conclude among other things that a 5-factor model is realistic and that the variation in H_2O, CO_2, UO_2 and the oxidation state of iron have been effected by near surface processes.

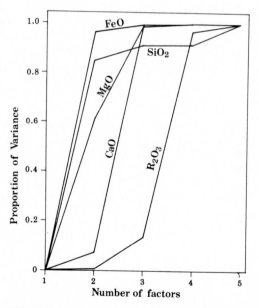

Fig. 8.4. A factor-variance diagram for the pyroxene data of Table 7.1 using the data from Table 8.6. When the number of factors is 5 or equal to the number of variables, the factor model exactly reproduces the original data and, therfore, accounts for all the variance in the variables. When the number of factors is 1, none of the variance in the variables is accounted for. A 3-factor model would be an acceptable choice as, except for R_2O_3, it accounts for the data reasonably well.

TABLE 8.6

Example of the data required to construct a factor-variance diagram. The differences between the estimated data values, using equation 8.35, and the actual values (from Table 7.1) are an indication of how well the model fits the data. The coefficients of determination are calculated using equation 8.36.

DATA FOR A 2-FACTOR MODEL

	Estimated data from equation 8.35, $\hat{x}_{ij}$					Differences $(x_{ij} - \hat{x}_{ij})$				
	SiO_2	R_2O_3	MgO	FeO	CaO	SiO_2	R_2O_3	MgO	FeO	CaO
1	55.41	4.21	0.90	25.58	13.91	2.34	−2.34	2.75	10.92	−13.68
2	49.59	4.02	26.80	10.51	9.08	0.38	0.18	1.84	5.24	−7.64
3	45.54	3.89	44.84	0.01	5.72	0.94	−2.56	2.37	3.52	−4.27
4	54.80	4.19	3.63	23.99	13.40	0.19	0.51	−2.04	−6.72	8.05
5	49.02	4.00	29.35	9.02	8.60	−0.30	−2.11	−2.50	−7.95	12.87
6	53.99	4.16	7.24	21.89	12.72	−1.14	0.86	−1.53	−5.41	7.22
7	50.26	4.04	23.82	12.24	9.64	−2.68	1.63	−2.09	−4.82	7.96
8	51.43	4.08	18.63	15.26	10.60	−1.32	0.59	0.27	1.07	−0.61
9	49.31	4.01	28.06	9.78	8.85	0.48	4.78	−1.73	−2.79	−0.75
10	52.19	4.10	15.22	17.24	11.24	0.87	−2.37	2.30	6.37	−7.16
11	49.34	4.01	27.91	9.86	8.87	1.17	−0.48	1.30	3.03	−5.01

Variances of original variables (Table 7.1)   Variances of differences, $var(x-x_j)$

11.00	4.92	180.28	99.31	73.74		1.91	4.91	4.37	37.81	68.60

Coefficient of determination, r_j^2

0.826	0.003	0.976	0.619	0.070

COEFFICIENTS OF DETERMINATION FOR ALL POSSIBLE FACTOR MODELS

Number of factors	SiO_2	R_2O_3	FeO	MgO	CaO
1	0.000	0.000	0.000	0.000	0.000
2	0.826	0.003	0.976	0.619	0.070
3	0.915	0.138	0.998	0.995	0.992
4	0.915	0.981	1.000	0.998	0.999
5	1.000	1.000	1.000	1.000	1.000

CHAPTER 9

MULTIPLE DISCRIMINANT ANALYSIS

9.1 INTRODUCTION

Petrological variation diagrams are frequently used to illustrate the *differences between* groups of analyses, often with diagrams that are also used to display the variation *within* the analyses. However, any diagram which tries to do both will always be a compromise. This is because principal components analysis, which is the best way to visualize the variation within data, is fundamentally different from multiple discriminant analysis, which is the best way to investigate the variation between groups of data. Before producing any variation diagrams, therefore, the petrologist must have a clear idea of why they are being produced. The fact that a particular type of diagram has been used before, *should not* be a sufficient reason.

The objective of multiple discriminant analysis is admirably summarized by Marriott (1974, p.32) and is "to examine how far it is possible to distinguish between members of various groups on the basis of observations made upon them". Note that discriminant analysis is *not* a method which will distinguish "natural" groups within sets of data (methods of attempting this are discussed in Chapter 10), as it relies upon prior knowledge of the grouping of the objects. The assignment of each set of observations to one of the two or more groups which are to be discriminated, must be done using criteria which do not include the variables to be used in the discriminant analysis, otherwise the results will be pre-judged. For example, if the major oxides are to be used as variables, it is quite in order to use discriminant analysis to distinguish between andesite and basalt analyses, if the rocks have been named on a basis of their mineralogy and petrography, but not if they have been named on a basis of their SiO_2 content. Fortunately, this problem is not too common, as in many instances the petrologist is interested in distinguishing groups from different tectonic settings or localities, such as different plutons or volcanic islands, so that the assignment into groups is done using locational data.

Once the observations have been assigned into groups, the multiple discriminant analysis can be performed. This gives results in terms of a series of linear functions called *discriminant functions* or eigenvectors, which can be thought of as directions in p-dimensional space which successively separate the groups from each other by decreasing amounts. Like principal components analysis, each eigenvector is associated with an eigenvalue. However, one major difference between the eigenvectors produced by principal components analysis and multiple discriminant analysis is that, in the case of multiple discriminant analysis, they are not generally at right angles to each other, although they are uncorrelated with each other (Blackith & Reyment, 1971; Tatsuoka, 1971). Like principal components analysis, multiple discriminant analysis is readily amenable to the production of plots, which makes it an attractive method for petrologists.

An extensive bibliography of 26 books and 547 research papers on the topic of discriminant analysis from a wide variety of disciplines is given by Cacoullos & Styan (1973).

9.2 GEOMETRIC INTERPRETATION

The geometric interpretation of discriminant analysis is illustrated in 2-dimensional space in Fig. 9.1, using two groups P and Q, which are shown by means of their probability contours as described in section 4.2. A discriminant analysis of such data would produce a single discriminant function or eigenvector, defining the normal to the "hyperplane" that best separates the two groups. One way of judging the efficiency of the separation of the two groups is to determine the amount of overlap of the distributions (the shaded areas) formed when the original data is projected onto the discriminant function as illustrated (see also section 9.5.3). Obviously, the amount of overlap would increase and, therefore, the effectiveness of the discrimination decrease, if the data were projected onto any other direction, such as x_1 or x_2.

In general, if there are g groups and p independent variables (one minus the actual number of variables if the data are closed), the number of discriminant functions will be the smaller of g - 1 and p. For example, with 9 oxides and 3 groups, there would be 2 discriminant functions, but with 12 groups there would only be 9 discriminant functions. This follows from the geometry of p-dimensional space. However, this does not mean that all of these discriminant functions will be significant or useful (see section 9.4). Furthermore, if the data are closed, or if several of the groups lie in a plane or along a line in p-dimensional space, some of the eigenvalues will have zero values, i.e. they will have no discriminating power. When more than two discriminant functions are obtained it is then possible to produce a plot showing the distribution of the groups in the discriminant space, by projecting the original data onto the succesive discriminant functions, in a manner directly analogous to that used for principal components analysis.

One of the basic assumptions of multiple discriminant analysis is that the variables are multivariate normally distributed and that the groups have equal dispersion matrices, i.e. the shape and orientation of the probabilty contours in Fig. 9.1 are the same for each group. As Marriott (1974) points out, however, departures from normality are not too serious, but unequal dispersion matrices can be a problem, as they theoretically cause the surface which best separates the groups to become curved. Although this can be dealt with by using

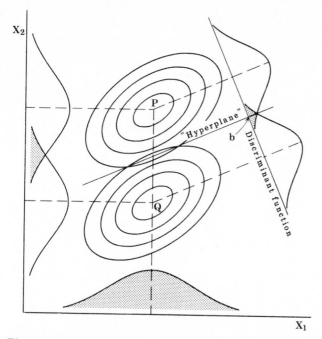

Fig. 9.1. The geometric interpretation of discriminant analysis using two groups P and Q, whose probability contours are shown for the two variables x_1 and x_2. The single discriminant function defines the normal to the "hyperplane" that best separates the two groups. The amount of misclassification can be judged by the amount of overlap of the distributions of the projected data (the shaded areas). Point b is the cut-off point on the discriminant coordinate scale.

quadratic discriminant functions, Marriott (1974) suggests the most practical way of handling the problem is to transform the data so that all the dispersion matrices become similar. However, if conventional multiple discriminant analysis is used for groups with unequal dispersion matrices, it would be extremely difficult to obtain any other linear function of the variables that would better separate the groups (E. J. Williams, per. comm.).

Fig. 9.1 also illustrates another important point. In selecting the variables to be used in multiple discriminant analysis, it must not be assumed that variables which have similar means in each group, will not be useful in discriminating between the groups, as they may be strongly correlated with other variables. For example, in Fig. 9.1 the means of x_1 for groups P and Q are identical, yet the amount of discrimination given by using the variables x_1 and x_2 is much better than that given by using variable x_2 alone.

Some examples of three group discriminant analysis are shown in Fig. 9.2, to illustrate the ways in which the location of the discriminant functions are dependent upon the dispersion matrices of the groups. The discriminant functions were calculated using computed sets of multivariate normally

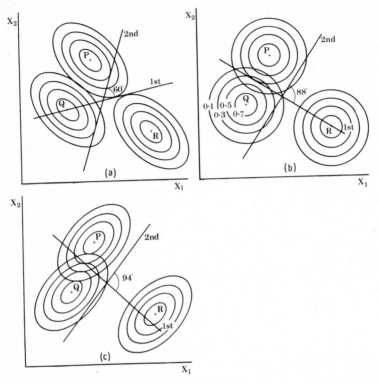

Fig. 9.2. Diagram to illustrate the relative locations of two discriminant functions in a 3 group discriminant analysis, using 2 variables x_1 and x_2. The discriminant functions were calculated using sets of multivariate normally distributed data produced by subroutine GGNSM from the IMSL library. In all the diagrams the variance of x_1 and x_2 are 1.0, but the covariances between x_1 and x_2 are as follows:- (a) -0.5; (b) 0.0 and (c) 0.5. The probability contours used in the diagrams (0.1, 0.3, 0.5 and 0.7) have been calculated theoretically for each dispersion matrix. Note that the 1st discriminant function is much more sensitive to the value of the dispersion matrix than the 2nd, and that none of the discriminant functions are orthogonal.

distributed data, produced by subroutine GGNSM from the IMSL library. The probability contours shown in Fig. 9.2 are calculated from the dispersion matrices input to subroutine GGNSM. In all three examples, the variance of x_1 and x_2 in each group is constant at 1.0, but the covariance varies from -0.5 in Fig. 9.2a, to 0.0 in Fig. 9.2b, and 0.5 in Fig. 9.2c.

As can be seen, the location of the 1st discriminant function is very dependent upon the values in the dispersion matrix, but the 2nd is more stable. The other point to notice is that the angle between the two discriminant functions is not a right angle, except when there is no correlation between the two variables. However, in most plots involving discriminant coordinates, the two discriminant functions are usually plotted at right angles to each other, so that the distribution of the data is then distorted with respect to the distribution in the original space, although, apart from the additional work involved, there is no reason why plots should not be produced with the discriminant functions plotted at the correct angle to each other.

9.3 BASIC MATHEMATICS

The detailed mathematical theory of discriminant analysis is complex and will not be given here as excellent accounts can be found elsewhere, e.g. Blackith & Reyment (1971), Cooley & Lohnes (1971), Gnanadesikan (1977), Marriott (1974) and Tatsuoka (1971). Instead, only the basic framework will be presented so that the reader will be aware of the general procedure.

The method involves calculating three pxp *sums of squares and products matrices*, which are denoted **T** for *total*, **A** for *among-groups* (also refered to by some authors as **B**, for *between-groups*) and **W** for *within-groups*. The individual terms of **T**, denoted t_{ij}, are given by:-

$$t_{ij} = \sum_{k=1}^{N} (x_{ik} - \bar{x}_i)(x_{jk} - \bar{x}_j) \tag{9.1}$$

where N is the total number of observations used in the discriminant analysis, $\bar{x}_i$ is the total sample mean of the ith variable, and x_{ik} is the value of the ith variable of the kth observation in the total sample. These terms are the same as those of the variance-covariance matrix of equation 7.1, multiplied by $N - 1$. Similarly, the individual terms of **A** and **W** are given by:-

$$a_{ij} = \sum_{k=1}^{g} N_k (\bar{x}_{ik} - \bar{x}_i)(\bar{x}_{jk} - \bar{x}_j) \tag{9.2}$$

and

$$w_{ij} = \sum_{k=1}^{g} \sum_{m=1}^{N_k} (x_{imk} - \bar{x}_{ik})(x_{jmk} - \bar{x}_{jk}) \tag{9.3}$$

where g is the number of groups being discriminated, N_k is the number of observations in the kth group ($N = \sum N_k$), x_{imk} is the ith variable of the mth observation in the kth group, and $\bar{x}_{ik}$ is the mean value of the ith variable in the kth group. In practice, only two of the matrices are usually calculated directly, the remaining one being found by difference using the relationship that:-

$$\mathbf{T} = \mathbf{A} + \mathbf{W} \tag{9.4}$$

The discriminant functions, which maximize the ratio of the among-groups to within-groups mean sum of squares, are then given by the eigenvectors of the matrix equation:-

$$(\mathbf{A} - \lambda_i \mathbf{W})\mathbf{v}_i = 0 \tag{9.5}$$

where λ_i is the ith eigenvalue and $\mathbf{v}_i$ is the ith eigenvector. Rewritten in the form:-

$$(\mathbf{W}^{-1}\mathbf{A} - \lambda_i\mathbf{I})\mathbf{v}_i = 0 \tag{9.6}$$

where $\mathbf{I}$ is an identity matrix, it can be seen that the required eigenvalues and eigenvectors are those of the matrix product $\mathbf{W}^{-1}\mathbf{A}$ (see equation 7.3). As long as $\mathbf{W}$ is not a singular matrix, the eigenvectors can then be extracted using one of the many subroutines that are available e.g. DIRNM (Cooley & Lohnes, 1971, p.198). If $\mathbf{W}$ is a singular matrix, however, special measures have to be taken (see section 9.3.1). Note that although $\mathbf{W}^{-1}$ and $\mathbf{A}$ are both symmetrical matrices, their product is not, so that it is not possible to utilize the same subroutines that are used to extract the eigenvectors from the symmetrical matrices of principal components analysis.

Once the discriminant functions have been found, *discriminant coordinates* (also referred to as *discriminant scores*) can be calculated for each set of observations, in the same way that principal component coordinates are calculated for principal components analysis, i.e. using equation 7.13. These give the location of the data points when projected onto each discriminant function and are given by:-

$$z_{jk} = \sum_{i=1}^{p} v_{ki}x_{ji} \tag{9.7}$$

where z_{jk} is the kth discriminant coordinate for the jth set of observations, v_{ki} is the ith term of the kth discriminant function, and x_{ji} is the ith variable of the jth observation.

If more than two discriminant functions exist, a diagram showing the relationships between the groups can then be produced by plotting the discriminant coordinates against each other as shown in Fig. 9.3.

Unlike principal components analysis, however, the discriminant functions or eigenvectors are not generally at right angles to each other. This can be checked by using equation 5.10 to calculate $\cos\theta_{jk}$, the cosine of the angle between the jth and kth eigenvectors, as:-

$$\cos\theta_{jk} = \sum_{i=1}^{p} v_{ki}v_{ji} \tag{9.8}$$

which, in general, will not be zero.

Numerical example

For an example of multiple discriminant analysis, we will use the raw values of normative Or, Ab and An of 13 basalts, 25 trachybasalts, 17 trachyandesites and 18 trachytes taken from the South Atlantic islands of Gough Island (Le Maitre, 1962), Tristan da Cunha (Baker et al., 1964) and St.

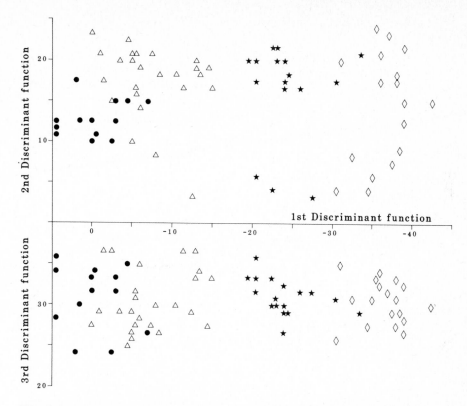

Fig. 9.3. Plot of the 2nd and 3rd discriminant coordinates against the 1st discriminant coordinate of Table 9.2, to show the relative distribution of the data points. The basalts are respresented by the symbol ●, the trachybasalts by △, the trachyandesites by ★ and the trachytes by ◊. Note that the 2nd and 3rd discriminant functions have virtually no ability to separate the groups.

Helena (Baker, 1969). The means and standard deviations of the four groups are given in Table 9.1, together with the total-, among- and within-groups sums of squares and products matrices. As there are 4 groups and 3 variables, there are 3 discriminant functions, which are also given. To emphasize the fact that the discriminant functions are not generally orthogonal, the angles between them have also been calculated using equation 9.8. As the data are not closed (the Or, Ab and An have not been recalculated to 100%), the inverse of the within-groups matrix $\mathbf{W}^{-1}$ and the matrix product $\mathbf{W}^{-1}\mathbf{A}$ are also given. Note that the sum of the 3 eigenvalues is equal to the sum of the diagonal terms of $\mathbf{W}^{-1}\mathbf{A}$ (see equation 7.4).

The interpretation of the 1st discriminant function is as one might expect; the four groups are separated mainly by decreasing Or and Ab and increasing An, i.e. the alkali feldspar to basic plagioclase ratio. As there are 3 discriminant functions, the 1st & 2nd and 1st & 3rd discriminant coordinates have been plotted against each other in Fig. 9.3, to show the relative locations

TABLE 9.1

Results of multiple discriminant analysis using the raw values of normative Or, Ab and An for four groups of rocks taken from the three South Atlantic islands of Gough Island (Le Maitre, 1962), Tristan da Cunha (Baker et al., 1964) and St. Helena (Baker, 1969). For further details see text.

Groups	Number	Means			Standard Deviations		
		Or	Ab	An	Or	Ab	An
Basalts	13	10.19	21.50	22.95	2.90	5.55	3.50
Trachybasalts	25	17.74	21.83	19.93	3.61	8.80	3.00
Trachyandesites	17	26.94	36.44	11.87	5.02	8.19	2.80
Trachytes	18	32.39	48.08	4.82	5.51	5.11	2.95
Total	73	22.15	31.65	14.86	9.10	13.30	7.54

Total-groups, $\mathbf{T}$

	Or	Ab	An
Or	5956.294	5079.537	-3953.750
Ab	5079.537	12743.927	-5722.122
An	-3953.750	-5722.122	4098.534

Among-groups, $\mathbf{A}$

	Or	Ab	An
Or	4624.252	6079.983	-3912.780
Ab	6079.983	8997.193	-5525.995
An	-3912.780	-5525.995	3462.031

Within-groups, $\mathbf{W}$

	Or	Ab	An
Or	1332.042	-1000.446	-40.971
Ab	-1000.446	3746.734	-196.128
An	-40.971	-196.128	636.503

Inverse Within-groups, $\mathbf{W}^{-1}$ (x 10^6)

	Or	Ab	An
Or	951.482	261.486	141.818
Ab	261.486	343.137	122.563
An	141.818	122.563	1617.979

Matrix product $\mathbf{W}^{-1}\mathbf{A}$

	Or	Ab	An
Or	5.4348	7.3540	-4.6769
Ab	2.8159	3.9998	-2.4950
An	-4.9298	-6.9760	4.3693

Discriminant functions
(eigenvectors of $\mathbf{W}^{-1}\mathbf{A}$)

	1st	2nd	3rd
Or	-0.683	0.830	0.301
Ab	-0.362	-0.302	0.344
An	0.634	0.468	0.890

Angles between discriminant functions

Pair	Cosine	Angle
1 & 2	-0.1609	99.3
1 & 3	0.2341	76.5
2 & 3	0.5625	55.8

Eigenvalues

1st	2nd	3rd
13.6730	0.1241	0.0069

of each data point. The discriminant axes, however, have been plotted as if they were orthogonal. Visual inspection of the data points shows that most of the discrimination is contained in the 1st discriminant function, which is in accord with the relative magnitude of the eigenvalues (see also section 9.4.2).

9.3.1 The problem of singular matrices

If **W** and **A** are singular matrices, as is the case with closed petrological data, W^{-1} the inverse of **W** will not exist. However, this does not mean that the matrix equation 9.6 cannot be solved. It simply means that there is an indeterminacy in the system which gives rise to an infinite number of solutions, all of which have the same discriminating power. In practice a subroutine such as DIRNM (Cooley & Lohnes, 1971, p.198) may give one of the infinite solutions when given singular matrices, if rounding-off errors within the computation determine the pth eigenvalue of **W** as a very small number, rather than its true value of zero. Unfortunately, the algorithm used in subroutine DIRNM does not work if one of the eigenvalues of **W** is zero.

One way to always obtain a solution is to delete one of the p variables from the discriminant analysis, so that **W** and **A** are no longer singular. The matrix equation can then be solved in the usual way, to give a set of discriminant functions involving p - 1 terms each. If we assume that the pth variable was omitted, using equation 9.7 we can write the discriminant coordinates as:-

$$z_{jk} = \sum_{i=1}^{p-1} v_{ki}x_{ji} \qquad (9.9)$$

If we set $v_{kp} = 0.0$, this can be rewritten as:-

$$z_{jk} = \sum_{i=1}^{p} v_{ki}x_{ji} \qquad (9.10)$$

Let us now subject each term of the discriminant functions to a linear transformation, so that we obtain new sets of unstandardized discriminant functions given by the equation:-

$$v_{ki}' = av_{ki} + b \qquad (9.11)$$

where a and b are constants. We can then write the new discriminant coordinates as:-

$$\begin{aligned}
z_{jk}' &= \sum v_{ki}'x_{ji} \\
&= \sum (av_{ki} + b)x_{ji} \\
&= a\sum v_{ki}x_{ji} + b\sum x_{ji} \\
&= az_{jk} + 100b
\end{aligned} \qquad (9.12)$$

where the summations are over i, and using the fact that for closed data $\sum x_{ji} = 100$. It, therefore, follows that the new set of discriminant coordinates have the same discriminating power as the original set, as they are linearly related – the constant terms, a and 100b, simply change the scale of measurement and shift the origin, respectively. Obviously, if $a = 1$ and $b = v_{kr}$, the rth term of the discriminant function, v_{kr}', will be zero, so that this is the solution obtained when the rth variable is omitted.

The only remaining problem is to decide which of the infinite number of solutions to use. From a purely practical point of view it is immaterial, but from the mathematical point of view there is a certain satisfaction about choosing the solution which gives the discriminant functions lying in the $p - 1$ hyperplane of constant composition with all the data. Le Maitre (1968) showed that the unstandardized discriminant functions given by:-

$$v_{ki}' = pv_{ki} - \sum v_{ki} \tag{9.13}$$

fulfil this condition, as the sum of their terms is zero (equation 5.11). As the form of this equation is identical to equation 9.11, it also follows that these discriminant functions will have the same discriminating power as those given by equation 9.10. Standardizing these discriminant functions will also have no effect upon the discriminating power as it is simply another linear transformation.

An alternative and more elegant solution is to extract the eigenvectors from the matrix product $\mathbf{W^-A}$, where $\mathbf{W^-}$ is the symmetrical generalized inverse matrix to be described in section 10.3.4. This may be done by using one of the standard subroutines for extracting eigenvectors from non-symmetrical matrices, such as EIGFR in the IMSL library. The resulting discriminant functions are the same as those obtained from equation 9.13, except that they are already standardized.

Numerical example

As an example of multiple discriminant analysis with closed data, we will use the values of Or, Ab and An from the previous numerical example recalculated to sum to 100%. The relevant data and matrices are shown in Table 9.2. As the number of independent variables is now 2, instead of 3, there will only be two discriminant functions. These may be calculated by omitting one of the variables, say An, from the calculation which would then give the eigenvectors in terms of Or and Ab only, as shown in Table 9.2. For example, taking the 1st discriminant function, we have:-

$$\sum v_{1i} = 0.799 + 0.601 = 1.400$$

Using equation 9.13, we can then calculate the terms of the discriminant function lying in the $p - 1$ hyperplane of constant composition as:-

$$v_{1Or} = 3 \times 0.799 - 1.400 = 0.997$$
$$v_{1Ab} = 3 \times 0.601 - 1.400 = 0.403$$
$$v_{1An} = 3 \times 0.0 - 1.400 = -1.400$$

TABLE 9.2

Results of multiple discriminant analysis using singular matrices. The data are the same as those used for Table 9.1, except that the values of normative Or, Ab and An were recalculated to sum to 100% before use.

Groups	Number	Means			Standard Deviations		
		Or	Ab	An	Or	Ab	An
Basalts	13	18.97	38.70	42.33	6.12	6.97	5.98
Trachybasalts	25	30.37	35.53	34.10	7.18	11.01	6.55
Trachyandesites	17	36.08	48.08	15.84	7.76	8.60	3.75
Trachytes	18	37.86	56.57	5.57	5.45	7.15	3.30
Total	73	31.51	44.21	24.28	9.36	12.26	14.92

Total-groups, **T**

	Or	Ab	An
Or	6312.300	-556.217	-5756.003
Ab	-556.217	10825.785	-10269.929
An	-5756.003	-10269.929	16026.214

Among-groups, **A**

	Or	Ab	An
Or	3154.923	-3414.198	259.459
Ab	-3414.198	5541.287	-2127.093
An	259.459	-2127.093	1867.456

Within-groups, **W**

	Or	Ab	An
Or	3154.923	-3414.198	259.459
Ab	-3414.198	5541.287	-2127.093
An	259.459	-2127.093	1867.456

Generalized Inverse **W**$^-$ (x 10^6)

	Or	Ab	An
Or	222.483	-6.124	-216.366
Ab	-6.124	85.909	-79.786
An	-216.366	-79.786	296.161

Matrix product **W**$^-$**A**

	Or	Ab	An
Or	1.9865	2.3653	-4.3519
Ab	0.7061	1.0862	-1.7924
An	-2.6927	-3.4516	6.1445

Discriminant functions (eigenvectors of **W**$^-$**A**)

	1st	2nd
Or	0.565	0.735
Ab	0.228	-0.676
An	-0.793	-0.059

Angles between discriminant functions

Pair	Cosine	Angle
1 & 2	0.3079	72.1

Eigenvalues

1st	2nd
9.0536	0.1615

Discriminant functions obtained by omitting An

	1st	2nd
Or	0.799	0.790
Ab	0.601	-0.613

which, when standardized so that their sum of squares is 1, give the 1st discriminant function as 0.565, 0.228 and -0.793. This is the same as the result obtained by extracting the eigenvectors and eigenvalues directly from the matrix product $\mathbf{W^-A}$ (Table 9.2). Note that the sum of the terms of each eigenvector is also 1, indicating that the discriminant functions all lie in the p - 1 hyperplane of constant composition.

9.4 TESTING THE SIGNIFICANCE OF THE DISCRIMINATION

Testing the significance of the discrimination can be approached in two different ways. One method is to produce an overall test for the differences between the group means, while the other is to see how many of the discriminant functions are significant. If only one discriminant function exists, both tests are equivalent.

9.4.1 *An overall test for differences between group means*

One commonly used method is a multivariate extension of the analysis of variance for testing the differences between several group means and is based on a criterion known as Wilks' Lambda, which is usually expressed as the ratio of the determinants of $\mathbf{W}$ to $\mathbf{T}$ (e.g Cooley & Lohnes, 1971, p.227; Tatsuoka, 1971, p.88). However, if the data are closed these two determinants are zero, so that Wilks' Lambda is indeterminate and cannot be calculated using this ratio. However, with multiple discriminant analysis, Wilks' Lambda can also be expressed in terms of the non-zero eigenvalues of the matrix $\mathbf{W^{-1}A}$ as:-

$$1/\Lambda = (1 + \lambda_1)(1 + \lambda_2) \, \dots \, \dots \, (1 + \lambda_r) \tag{9.14}$$

where Λ is Wilks' Lambda and $\lambda_1, \lambda_2, \dots \lambda_r$ are the non-zero eigenvalues. The value of Wilks' Lambda can then be incorporated into Rao's R-statistic which is given by:-

$$R = \frac{1 - \Lambda^{1/s}}{\Lambda^{1/s}} \frac{n_2}{n_1} \tag{9.15}$$

where

$$s = \sqrt{\frac{p^2(g-1)^2 - 4}{p^2 + (g-1)^2 - 5}} \tag{9.16}$$

$$n_1 = p(g-1) \tag{9.17}$$

$$n_2 = s\left[(N-1) - (p+g)/2\right] - p(g-1)/2 + 1 \tag{9.18}$$

where p is the number of independent variables (one minus the actual number of variables if the data are closed), g the number of groups being discriminated, and N the total number of observations. As R is distributed approximately as an F-variate with n_1 and n_2 d.f. (or exactly if either p = 1 or 2, or g = 2 or 3), it can be tested for signficance as shown in section 2.10, i.e.

if R exceeds the critical value of F, given in Table A4 in the Appendix, we can then state that there is a significant difference between the group means and that the discriminant analysis is effective.

Numerical example

Using the data of Table 9.1, we will now test the significance of the overall differences between the group means of the basalts, trachybasalts, trachyandesites and trachytes. Using equations 9.16, 9.17 and 9.18 we obtain:-

$$s = \sqrt{\left[\left\{3^2(4-1)^2 - 4\right\}/\left\{3^2 + (4-1)^2 - 5\right\}\right]} = \sqrt{(77/13)} = 2.4337$$

and $1/s = 0.4109$

$$n_1 = 3(4-1) = 9$$

$$n_2 = 2.4337\left[73 - 1 - (3+4)/2\right] - 3(4-1)/2 + 1$$

$$= 166.7085 - 3.5 = 163.2085$$

Then using equation 9.14 we obtain:-

$$1/\Lambda = (1 + 13.6730)(1 + 0.1241)(1 + 0.0069) = 16.6077$$

which gives $\Lambda = 0.060213$ and $\Lambda^{1/s} = 0.3152$, so that using equation 9.8 we finally obtain:-

$$R = \frac{(1 - 0.3152) \times 163.2085}{(0.3152 \times 9)} = 39.4 \quad \text{with 9 and 163 d.f.}$$

As would be expected, the differences between the 4 group means are highly significant, as the critical value of the F-distribution in Table A4 in the Appendix is less than 2.6 at the 99% confidence level.

9.4.2 Testing successive discriminant functions

One of the problems with multiple discriminant analysis is to decide how many of the discriminant functions contribute significantly to the discrimination. This can done using a chi-square test (Cooley & Lohnes, 1971, p.249; Tatsuoka, 1971, p.164) which determines the amount of discrimination contained in the $r - k$ discriminant functions, after the first k have been accepted as significant. The chi-square statistic is computed as follows:-

$$\chi^2 = -\left[N - (p+g)/2 - 1\right] \log_e \Lambda' \tag{9.19}$$

where

$$1/\Lambda' = (1 + \lambda_{k+1})(1 + \lambda_{k+2}) \dots \dots (1 + \lambda_r) \tag{9.20}$$

and has $(p-k)(g-k-1)$ d.f. The value of χ^2 can then be tested for significance using Table A5 in the Appendix. When $k = 0$, this test should give the same result as Rao's R-statistic used in the preceding section.

Numerical example

Again using the data of Table 9.1, we will see if the 2nd discriminant function is significant. With k = 1, equation 9.20 gives:-

$$1/\Lambda' = (1 + 0.1241)(1 + 0.0069) = 1.1319 \quad \text{and} \quad \Lambda' = 0.8835$$

Equation 9.19 then gives:-

$$\chi^2 = -(73 - 3.5 - 1)\log_e 0.8835 = 8.48$$

with $(3 - 1)(4 - 1 - 1) = 4$ d.f.

At the 95% confidence level, the critical value of χ^2 in Table A5 in the Appendix is found to be 9.49, so that there is no evidence to suggest that the 2nd discriminant function is significant. This can also been seen visually in Fig. 9.3. One of the reasons that only one discriminant function is significant in this example, is that the four groups are roughly collinear in space.

9.5 THE LOGIC OF ASSIGNMENT RULES

One of the objectives of discriminant analysis is to be able to classify unknown objects into one of the pre-defined groups. This is usually done by means of simple assignment rules which, in the case of two group discriminant analysis, are relatively easy to define; but with multiple groups the rules are less clear cut. It should be noted, however, that the rigorous application of these rules can be dangerous, unless it is known that the object belongs to one of the groups, as the method will assign any object to one of the groups, whether or not it actually belongs to any of the groups.

9.5.1 For two group discriminant analysis

In the case of two group discriminant analysis, the assignment rule is simple. In Fig. 9.1, for example, any object falling above the "hyperplane" would be assigned to group P, while any falling below would be assigned to group Q. In terms of discriminant coordinates, which are the values actually used, the assignment rule can be stated as follows:-

An unknown is assigned to the group whose mean is numerically
on the same side of the cut-off value as itself. (9.25)

where the *cut-off* value is the discriminant coordinate of any point on the "hyperplane", e.g. it is the discriminant coordinate of the cut-off point b in Fig. 9.1. Mathematically this may be expressed as:-

If $(z_u - z_b) / (z_P - z_b)$ is positive, then assign the unknown to group P (9.26)

where z_u, z_b and z_P are the discriminant coordinates of the unknown, point b (the cut-off value) and the mean of group P, respectively. The only problem is to define the cut-off value. If it is assumed that both groups have equal dispersion matrices (as illustrated), then the cut-off value is the mean of the discriminant coordinates of the means of groups P and Q. Note that it is independent of the number of observations in the two groups. If the dispersion matrices are not equal, however, then the problem is more difficult. One method is to produce actual sample frequency distributions of the discriminant coordinates of the two groups and to choose the cut-off value for which the proportion of misclassification is the same for both groups (Le Maitre, 1980). Unless the samples are large, however, this will produce a biased result.

9.5.2 *For three or more group discriminant analysis*

When more than two groups are involved, the assignment rules become difficult to generalize. For example, consider the three groups P, Q and R shown in Fig. 9.2, together with their two discriminant functions. If one applies the simple strategy of two group assignment rules to two discriminant functions one would finish up with 4 fields (2^2) into which an unknown could fall. With three discriminant functions there would be 8 fields (2^3) etc. Obviously, there will always be more fields than groups, so that many groups will occupy more than one field. The sort of strategy that has to be worked out, therefore, can only be done by inspection of the location of the group means in the individual plots. For example, in Fig. 9.2a the 1st discriminant function basically separates group Q from group R, while the 2nd separates P from Q. In Fig. 9.2b, however, the 1st discriminant function separates group R from the other two, while the 2nd separates P from Q. An alternative strategy is to simply draw boundaries on the plots between the groups, in which case the boundaries can be curved. This, of course, is the approach taken by many petrologists when using conventional variation diagrams to separate groups.

Another approach to the problem is to perform two group discriminant analysis on all pairs of groups, but this, however, can lead to some anomalies in the use of the assignment rules, as shown by Le Maitre (1980). It is also a very time-consuming exercise unless the relative locations of all the groups are known, in which case only adjacent pairs need be used.

9.5.3 *Use in estimating amount of misclassification*

If satisfactory assignment rules can be established, they may then be used to estimate the amount of misclassification within the samples used, which can be used as a measure of the effectiveness of the discriminant analysis. Unfortunately, unless large samples are used, the value obtained will be biased and is likely to be less than for other samples and for the population as a whole. If necessary, the bias can be removed by adopting the following procedure. Each object in turn is assigned to a particular group on assignment rules worked out from a discriminant analysis performed on the remaining objects in the sample. Such a procedure involves a considerable amount of calculation but, fortunately, it is only necessary for small samples.

Note that the amount of misclassification is a measure of the chance of an individual object being incorrectly assigned to a group, and depends only upon the overlap between the groups and not upon the number of observations in the samples of the groups. However, the two tests of section 9.4 are testing for significant differences between the group means, and depend upon the total number of observations in the groups. For a given overlap between groups, therefore, the more observations that are in the group samples, the more significant the differences between the group means will becomes.

9.6 COMPARISON WITH PRINCIPAL COMPONENTS ANALYSIS

To illustrate the fundamental differences between multiple discriminant analysis and principal components analysis, some data from three South Atlantic islands have been used. They consist of a total of 77 analyses, each recalulated to 100% of the nine major oxides, of which 27 are from Gough Island (Le Maitre, 1962; all Table 10 except G121), 31 from Tristan da Cunha (Baker et al., 1964; all Table 6 except 114 and 62.1) and 19 from St. Helena (Baker, 1969, all Table 2 except 804).

The 1st two eigenvectors and their eigenvalues obtained from *principal components analysis* of all the 77 analyses, using a variance-covariance matrix, are given in Table 9.3. As can be seen, the direction of the 1st eigenvector accounts for 88.8% of the total variance and is controlled mainly by decreasing SiO_2 with increasing FeO, MgO and CaO. In petrological terms, this eigenvector is maximizing the differences between the basic and acid ends of the rock series, so that when the 1st and 2nd principal component coordinates are plotted against each other a more or less conventional variation diagram is produced, as in Fig. 9.4a, where the basalts are to the left and the trachytes to the right of the diagram. It can also be seen that the analyses from Tristan da Cunha tend to plot at the bottom of the diagram, but that the St. Helena and Gough Island analyses are completely intermingled. Obviously, such a diagram would be of little use in discriminating analyses from each of the three islands.

The two discriminant functions obtained when *multiple discriminant analysis* is performed on the three groups are also shown in Table 9.3. The first thing to notice is that the discriminant functions are all at high angles to the principal components analysis eigenvectors. In other words, discriminant analysis is looking "end-on" at the differentiation trends in space and is, therefore, selecting a projection which will place the trachytes and basalts from each island on top of each other, in order to maximize the differences between the islands, as can clearly be seen in Fig. 9.4b. The separation into distinct groups is remarkably good, with little overlap. Obviously, such a clear separation would have been extremely difficult to predict from Fig. 9.4a. Inspection of the terms of the 1st and 2nd discriminant functions indicates that they are controlled mainly by K_2O and Na_2O, respectively, so that one can conclude from the location of the group means that, in general, the analyses from Tristan da Cunha are richer in K_2O than analyses from the other two islands, while St. Helena is enriched in Na_2O with respect to Gough Island.

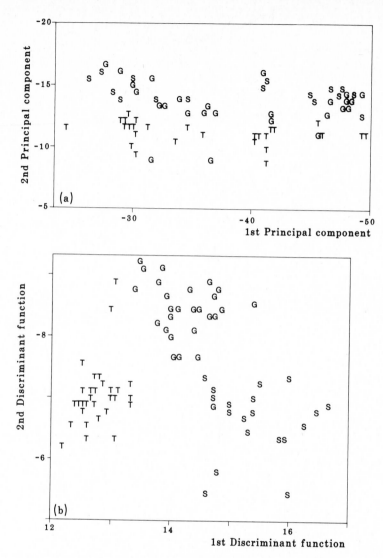

Fig. 9.4. Two diagrams to illustrate the difference in approach of multiple discriminant analysis and principal components analysis using data from Gough Island (G), Tristan da Cunha (T) and St. Helena (S) in the S. Atlantic. (a) a plot of the 1st versus 2nd principal component coordinates using the eigenvectors given in Table 9.3. This is similar to a conventional type of variation diagram with the basic rocks, which are to the left, being widely separated from the more acidic types, which are to the right. (b) a plot of the 1st versus 2nd discriminant coordinates using the discriminant functions of Table 9.3. Both the axes are at high angles to those in (a) so that one is looking "end-on" at the trends to maximize the differences between the rock series. The separation into distinct island groups is remarkably good.

TABLE 9.3

Results of multiple discriminant analysis and principal components analysis, using a variance-covariance matrix, on analyses from Gough Island, Tristan da Cunha and St. Helena in the S. Atlantic Ocean to illustrate the differences between the two methods. Note the large angles between the discriminant functions and the principal components eigenvectors.

	Means			Discriminant functions		Principal Components	
	Gough	Tristan	St.Helena	1st	2nd	1st	2nd
SiO_2	54.66	51.46	54.74	0.265	0.190	0.743	0.353
TiO_2	1.98	2.53	1.56	-0.006	0.304	-0.158	-0.083
Al_2O_3	17.73	18.44	17.57	0.079	0.015	0.135	-0.382
Fe_2O_3	3.07	3.23	2.92	0.264	0.109	-0.064	-0.370
FeO	5.19	4.99	6.32	0.406	0.049	-0.300	0.637
MgO	3.51	3.02	2.71	0.085	0.164	-0.300	0.291
CaO	5.28	7.16	4.99	-0.305	-0.223	-0.401	-0.190
Na_2O	4.48	5.04	6.11	-0.021	-0.850	0.171	-0.012
K_2O	4.10	4.14	3.08	-0.767	0.240	0.176	-0.244
		Eigenvalues as %		72.7	27.3	88.8	4.9

Angles and cosines of angles between vectors

	Discriminant functions			
	1st		2nd	
	cosine	angle	cosine	angle
1st Principal component	0.028	88.4	0.010	89.4
2nd Principal component	0.495	60.3	0.069	86.0
1st Discriminant function			0.014	89.2

9.7 APPLICATION TO METAMORPHIC PHASE DIAGRAMS

Although triangular diagrams, such as the ACF and AKF, first introduced by Eskola (1915), and the more recent AFM (Thompson, 1957), are commonly used by metamorphic petrologists to illustrate their mineral assemblages, they all have some disadvantages. For example, because of the type of projection used, some mineral phases cannot be represented which led Turner (1968, p.175) to state that "clearly the ACF diagram is a somewhat crude representation of mineral paragenesis". Similarly, both the ACF and AKF diagrams add FeO and MgO together, which in many cases is unrealistic. The AFM diagram does not suffer from this disadvantage, but it does suffer from the fact that it is a projection of a projection and that it "is not practicable for plotting total-rock compositions" (Turner, 1968, p.179). The major reason for these shortcomings is that the diagrams have been designed to reduce the dimensionality of the problem by ignoring certain mineral phases and/or grouping certain oxides together. As mentioned in section 5.7, triangular diagrams also suffer from the disadvantage that the lever rule (equation 5.16) is not generally applicable.

As the objective of these diagrams is to show the relationship of the mineral phases to each other and, if possible, to the bulk rock composition from which they came, it would seem to be logical to use multiple discriminant analysis to maximize the differences between all the mineral groups of interest. As long as more than two mineral groups are involved, a discriminant plot could then be produced (e.g. Fig. 9.5) which will have two major advantages over the conventional triangular plots. Firstly, *any* composition

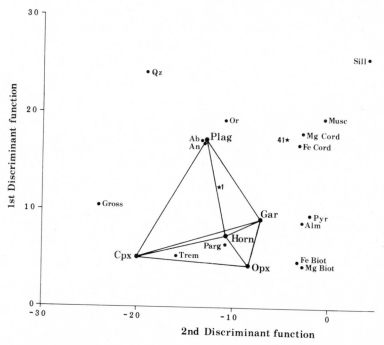

Fig. 9.5. A discriminant plot of data from Broken Hill, New South Wales, using the discriminant functions of Table 9.4. The direction of the axes has been chosen so that the orientation of the resulting diagram approximately corresponds with that of Fig. 9.6. The means of the five groups of minerals in Table 9.4 are plotted using the following abbreviations:- Cpx = clinopyroxene; Gar = garnet; Horn = hornblende; Opx = orthopyroxene and Plag = plagioclase. Other relevant end-member compositions taken from Table A13 in the Appendix, have also been plotted with the following abbreviations:- Ab = albite; Alm = almandine; An = anorthite; Biot = biotite; Cord = cordierite; Gros = grossular;   Musc = muscovite;   Or = orthoclase;   Parg = pargasite; Pyr = pyrope; Qz = quartz; Sill = sillimanite and Trem = tremolite. Mg and Fe refer to the respective end-members of the biotite and cordierite series. Two actual rock compositions (Binns, 1964) are also plotted as 1, a basic gneiss consisting of orthopyroxene, clinopyroxene, hornblende, plagioclase and quartz and 41, a pelitic schist consisting of garnet, biotite, muscovite, sillimanite and quartz. The tie-lines correspond with those shown in Fig. 9.6.

can be projected into the diagram to give a useful reference point and secondly, the lever rule is applicable as the discriminant plot is a linear projection from the original oxide space. Although this may seem a somewhat lengthy procedure to produce a plot, once the discriminant functions have been obtained, the amount of calculation involved is no more than that for the conventional triangular diagrams.

The method is illustrated with some analyses of hornblende, clinopyroxene, orthopyroxene, garnet and plagioclase, taken from the regionally metamorphosed basic rocks of Broken Hill, New South Wales (Binns, 1965a, 1965b). As the plagioclase analyses only record CaO, Na_2O and K_2O, the Al_2O_3 and SiO_2 were calculated assuming stoichiometry, and only those analyses which then summed to between 95% and 105% were accepted. After MnO was added to FeO each analysis was recalculated to 100% of the 10 major oxides as shown in Table 9.4. The first two discriminant functions obtained from a multiple discriminant analysis of these five groups are also given in Table 9.4,

TABLE 9.4

Some results of multiple discriminant analysis using 5 groups of metamorphic minerals from Broken Hill, New South Wales, including the group means, the first two discriminant functions and the discriminant coordinates of the means. Each analysis was recalculated to 100% after MnO had been added to FeO. Abbreviations used:- Horn = hornblende; Cpx = clinopyroxene; Opx = orthopyroxene; Gar = garnet and Plag = plagioclase.

		M e a n s				Discriminant functions	
	Horn.	Cpx.	Opx.	Gar.	Plag.	1st	2nd
SiO_2	42 90	50.23	49.14	36.49	50.46	0.241	-0.192
TiO_2	1.60	0.31	0.28	0.25	0.00	0.758	0.433
Al_2O_3	11.38	1.81	1.19	21.10	31.74	0.265	0.176
Fe_2O_3	2.65	0.73	0.46	0.81	0.00	-0.101	-0.150
FeO	19.00	16.36	34.87	31.81	0.00	-0.124	0.013
MgO	7.66	9.28	13.02	1.81	0.00	-0.265	0.070
CaO	11.07	20.93	0.96	7.73	14.41	-0.140	-0.547
Na_2O	1.36	0.32	0.07	0.00	3.30	-0.373	-0.281
K_2O	0.70	0.02	0.01	0.00	0.10	-0.064	-0.096
H_2O	1.68	0.00	0.00	0.00	0.00	-0.197	0.574

Discriminant coordinates of means

	1st	2nd
Horn.	7.5	-10.7
Cpx.	5.2	-20.0
Opx.	4.4	-8.4
Gar.	9.0	-7.0
Plag.	17.3	-12.9

as are the means of each mineral group, together with their first two discriminant coordinates. These five means have been plotted in Fig. 9.5 together with some end-member compositions of other minerals occurring in the Broken Hill rocks, using data from Table A13 in the Appendix. Two actual rock compositions have also been plotted using the analysis numbers of Binns (1964). They are a basic gneiss from zone A (Anal. 1, p.289) which consists essentially of orthopyroxene, clinopyroxene, hornblende, plagioclase and quartz, and a pelitic schist from zone C (Anal. 41, p.303) which consists essentially of garnet, biotite, muscovite, sillimanite and quartz. For comparative purposes, a conventional granulite facies ACF diagram for these rocks (based on Turner, 1968, p.327) is shown in Fig. 9.6, together with the location of the basic gneiss (1) and the pelitic schist (41). To further aid comparison the tie-lines in Fig. 9.6 have been drawn in Fig. 9.5.

There is good agreement between the two diagrams and the graphical interpretation of the discriminant plot is basically no different from that of the triangular diagram. The discriminant plot does, however, have several advantages over the triangular projections which can be summarized as follows:-

1. The minerals used in the multiple discriminant analysis will, in general, be more evenly spread. As these minerals are chosen by the petrologist as being of interest, this should aid their detailed graphical interpretation.
2. Any composition representable in terms of the oxides used in the multiple discriminant analysis can be projected into the discriminant plot. It, therefore, follows that tie-lines and/or compatability polygons can be drawn for *all* the mineral phases in the rock.

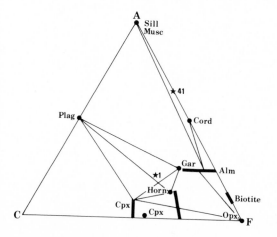

Fig. 9.6. A conventional ACF diagram for rocks of the granulite facies from Broken Hill, New South Wales based on a diagram from Turner (1968, p.327). The abbreviations are the same as those for Fig. 9.5. The two rock compositions plotted in Fig. 9.5 have also been plotted.

3. In general, minerals which plot in the same position in the triangular projection (e.g. muscovite and sillimanite, almandine and pyrope in the ACF diagram) will plot in different locations in the discriminant plot.

4. The lever rule is applicable, as the discriminant plot is a linear projection from the original oxide space. This enables a rough check to be made on any gross analytical or projection errors as, for example, a bulk rock composition must always plot within the external polygon formed by joining together all pairs of mineral phases occurring in the rock.

In all fairness, the use of the discriminant plot to represent metamorphic assemblages also has some relatively minor disadvantages. Firstly, multiple discriminant analysis must be performed on the mineral groups before any plot can be produced. In practice, however, this is a relatively easy with modern computer packages and is a task that need only be done once for each type of metamorphic assemblage. Secondly, several analyses of each mineral phase should be available. Ideally there should be more analyses of each phases than oxides being used so that, for example, if 10 oxides are being used, there should be more than 10 analyses of each phase. However, in practice it does not matter if some groups have fewer analyses than others, as long as the total number of analyses is greater than the number of oxides being used multiplied by the number of groups. If actual mineral analyses are not available, the multiple discriminant analysis can be performed using a randomly generated data set, as long as the means and variance-covariance matrices of each group are known or assumed. The final judgement as to the use of such discriminant plots, of course, must be left to the petrologist, but the simplicity of the approach has much to recommend it, when compared with some of the more elaborate projections in use today.

CHAPTER 10

CLUSTER ANALYSIS AND CLASSIFICATION

10.1 INTRODUCTION

This chapter deals with two methods that can be of help in classifying unknown objects. In the first, called *cluster analysis*, no assumptions are made about the data and the method seeks to locate any "natural" groupings or clusters. If satisfactory clusters can be located, these may be used as a basis for constructing a classification by using either discriminant analysis or the second method, which has been called *classification procedure*. This latter method can be used to classify unknowns when the characteristics of the groups are already known. It is, therefore, similar to discriminant analysis, but it has the major advantage that it can tell if an unknown is unlikely to belong to any of the predefined groups.

10.2 CLUSTER ANALYSIS

Cluster analysis is not a single technique but a series of procedures, with the common aim of locating "natural" groupings or clusters in the data. This is usually achieved by using one of several simple algorithms to form the groups. So far in petrology, the technique has not been widely used, unlike the field of numerical taxonomy, where it has been used extensively to divide sets of organisms into groups, often with a view to providing a basis for a classification. Excellent descriptions of some of the methods and problems encountered in cluster analysis can be found in Kendall (1973) and Hartigan (1975), who also includes computer programs for most of the algorithms used in cluster analysis.

The algorithms which are used to form the groups can be either divisive or agglomerative. A *divisive algorithm* initially regards all the data as a single group, which it continually subdivides according to some specified criterion, until each goup contains only one object. The criterion used usually involves maximizing or minimizing a sum of squares of some sort. However, these methods are seldom used due to the enormous amount of computer time required, for example, Gower (1967) estimated that one method would take more than 54,000 years for 41 objects! In practice, therefore, *agglomerative algorithms* are nearly always used and the rest of the discussion will be confined to these methods. They start with each object as an individual group and continually join the most similar objects and groups together, using a particular set of rules called a *linkage method*, until a single group has been formed. For both methods the final grouping is usually illustrated by means of a *dendrogram* or *tree-diagram*, e.g. Figs. 10.1b, 10.1c and 10.1d. Blackith & Reyment (1971), Gower (1967), Hartigan (1975) and Marriott (1974) further discuss the use of these methods.

Before a cluster analysis can be performed, therefore, the user has to answer two questions. Firstly, how is the similarity of objects to be measured and secondly, what type of linkage method is to be used?

10.2.1 The choice of similarity measure

The similarity between two objects is expressed numerically by a quantity called a *similarity measure*, or by some authors, a *dissimilarity measure*. Many types have been proposed and Wishart (1969) gives a convenient tabulation of some of them. Many of these, however, involve binary variables, e.g. 1 for present, 0 for absent, and because of their restricted use in petrology they will not be discussed further. Of those similarity measures which use real variables the following three are in common use.

1. The *Euclidean distance* measured between the ith and jth object, d_{ij}, which is given by:-

$$d_{ij} = \sqrt{\sum_{k=1}^{p} (x_{ik} - x_{jk})^2} \qquad (10.1)$$

where x_{ik} and x_{jk} are the kth variables of the ith and jth object, respectively, which is simply a restatement of equation 5.5. The value of d_{ij} will be zero for perfect similarity and will become more positive as the objects become more dissimilar. Some authors divide d_{ij} by p, the number of variables, but this seems somewhat pointless. Similarly, to save computation time, d_{ij}^2 could be used instead of d_{ij}.

2. The *correlation coefficient* between the ith and jth object, r_{ij}, given by:-

$$r_{ij} = \frac{\sum (x_{ik} - \bar{x}_i)(x_{jk} - \bar{x}_j)}{\sqrt{\sum (x_{ik} - \bar{x}_i)^2 \sum (x_{jk} - \bar{x}_j)^2}} \qquad (10.2)$$

where $\bar{x}_i$ and $\bar{x}_j$ are the means of the variables of the ith and jth object, respectively, and the summations are over k. Although this is similar to equation 4.6, notice that the roles of variable and object have been interchanged. The values of r_{ij} will range from 1.0 for perfect similarity to -1.0 for maximum dissimilarity.

3. The *cosine theta coefficient* which measures the cosine of the angle subtended by the ith and jth objects at the origin, and is given by:-

$$\cos \theta_{ij} = \sum_{k=1}^{p} x_{ik}x_{jk} \left/ \left[\sum_{k=1}^{p} x_{ik}^2 \sum_{k=1}^{p} x_{jk}^2 \right]^{1/2} \right. \qquad (10.3)$$

This is the same measure as the coefficient of proportionality used in factor analysis (equation 8.25). The values of $\cos \theta_{ij}$ will range from 1.0 for perfect similarity to 0.0 for maximum dissimilarity if all the values are of the same sign, or -1.0 if they have different signs.

Unfortunately, there is no right or wrong choice of similarity measure for a particular set of data, as to quote Marriott (1974, p.53) "a measure of dissimilarity is essentially a subjective measure, and the experimenter must take the responsibility for deciding what characters to include, and what weights to give them". This is similar to the choice that has to be made in principal components analysis (see section 7.3).

However, a few guidelines can be given to help in the choice. For example, the Euclidean distance is scale dependant and hence will be weighted in favour of those variables with large numerical values. As a result it is often recommended that each variable is standardized to have zero mean and unit variance before use, by using equation 1.2. This, of course, gives equal weights to all the variables, which the user may feel is not warranted. This led Ewart & Le Maitre (1980) to use a logarithmic transformation before calculating the Euclidean distances to be used in a cluster analysis of data involving modes and major oxides as percentages and minor elements as ppm. The Euclidean distance does, however, have the useful property that it is invariant under a rigid rotation or translation of axes. Furthermore, Davis (1973) suggests that it is better able to represent data in which a few extreme values are present, as the values of d_{ij} are not restricted to lie between definite bounds, such as 1.0 and -1.0. He also suggests that the Euclidean distances are less susceptible to changes in the linkage method used. In taxonomy, however, where the shape of objects is often more important than their size, the correlation and cosine theta coefficients are often used. As these two coefficients are independent of scale, there is no need to standardize the data before use. If, however, the variables *within* each object are standardized to have zero mean and unit variance, the two coefficients are equal, as $\bar{x}_i$ and $\bar{x}_j$ in equation 10.2 become zero. The relationships of these three measures to change of scale, translation of origin and rotation of axes, are summarized in Table 10.1. For further discussion on the advantages and disadvantages of some of the similarity measures, the reader may also refer to Davis (1973), Gower (1967) and Marriott (1974).

TABLE 10.1

Summary of the behaviour of three similarity measures under scale changes, translation of axes and rigid rotation of axes.

	Scale change	Translation	Rigid rotation
Euclidean distance	Dependent	Independent	Independent
Correlation coefficient	Independent	Independent	Independent
Cosine theta coefficient	Independent	Dependent	Independent

Having decided which measure to use, a similarity measure has to be calculated for all possible pairs of objects. These values form the *similarity matrix*, which is of size nxn, where n is the number of objects. As this matrix has to be examined every time a new linkage is made, it is more efficient of computer time if it is able to be stored in the central memory or core of the computer, which limits the number of objects that can be processed to about 200 to 300, unless a computer with a large virtual memory is being used. Fortunately, the similarity matrix is symmetrical, as $d_{ij} = d_{ji}$, so that only the upper or lower half of the matrix need be stored. However, this still means that for 200 objects an array of size $(200x199)/2 = 19900$ has to be allocated.

10.2.2 *The choice of linkage method*

Of equal importance to the choice of similarity measure, is the choice of linkage method to be used. Although many methods have been suggested (Hartigan, 1975), only three appear to be in common use in the geological sciences. They are:-

1. The *single-linkage method*, in which an object is connected to a group if it has the highest similarity with *any individual* object already in the group. As the only manipulation involved after the similarity matrix has been computed is ranking, the linkages can be performed by simple inspection of the matrix. It is also the only method that is easy to depict graphically. Consider the 2-dimensional example of the 7 objects A to G in Fig. 10.1a. The coordinates of the objects are given in Table 10.2, together with two similarity matrices, one based on the Euclidean distance and the other on the cosine theta coefficient - the correlation coefficient cannot be used in this example, as with only two variables its value is always 1. Let us now consider how the single-linkage method works, using the Euclidean distance as a similarity measure. Inspection of Fig. 10.1a and the similarity matrix shows that A and B are the two closest objects (0.54), so that they form the 1st link. The next shortest distance is between C and B (0.80), so that C is joined to group AB. However, the next shortest distance of 1.12 between A and C is ignored, as these two are already in the same group. The 3rd link is, therefore, between E and F (1.25), and starts a new group. The 4th linkage is then between F and G (1.44), the 5th between A and D (1.80) and finally, the two groups ABCD and EFG are joined between A and F (1.94). All these linkages are numbered in Fig. 10.1a and are shown as a dendrogram in Fig. 10.1b, in which the vertical scale is the similarity measure, in this case the Euclidean distance. As an exercise, the reader may care to confirm that the similarity matrix based on the cosine theta coefficient in Table 10.2 will produce the dendrogram shown in Fig. 10.1c - remember that with this measure the larger the value (more positive) the more similar the objects. Notice how in this example, groupings are strongly dependent upon the similarity measure used, which is understandable bearing in mind what the measures represent.

TABLE 10.2

Similarity matrices based on the Euclidean distance and cosine theta coefficient for a set of hypothetical data consisting of 2 variables and 7 objects. Note that as the similarity matrices are symmetrical, only the top or bottom halves need be stored in a computer.

Raw data

	A	B	C	D	E	F	G
x_1	1.20	1.00	0.20	0.20	2.20	2.80	4.00
x_2	1.50	1.00	1.00	3.00	3.70	2.60	1.80

Similarity matrix based on Euclidean distances

	A	B	C	D	E	F	G
A	0.00	0.54	1.12	1.80	2.42	1.94	2.82
B	0.54	0.00	0.80	2.15	2.95	2.41	3.10
C	1.12	0.80	0.00	2.00	3.36	3.05	3.88
D	1.80	2.15	2.00	0.00	2.12	2.63	3.98
E	2.42	2.95	3.36	2.12	0.00	1.25	2.62
F	1.94	2.41	3.05	2.63	1.25	0.00	1.44
G	2.82	3.10	3.88	3.98	2.62	1.44	0.00

Similarity matrix based on cosine theta coefficient

	A	B	C	D	E	F	G
A	1.000	0.994	0.888	0.821	0.990	0.989	0.890
B	0.994	1.000	0.832	0.753	0.969	0.999	0.935
C	0.888	0.832	1.000	0.991	0.943	0.811	0.581
D	0.821	0.753	0.991	1.000	0.892	0.728	0.470
E	0.990	0.969	0.943	0.892	1.000	0.959	0.819
F	0.989	0.999	0.811	0.728	0.959	1.000	0.947
G	0.890	0.935	0.581	0.470	0.819	0.947	1.000

2. The *unweighted average method.* In this an object is linked to a group if it has the highest similarity with the average similarity measure of the group. Once an object has been linked to a group a new average is computed for the group, by summing all the original similarity measures of the objects in the group and dividing by the number of objects in the group, i.e. the group average is the unweighted average of all objects in the group. This method has the property that an object incorporated into a large group has little influence on the average similarity measure of the group.

3. The *weighted-pair group average method.* This is similar to the previous method, except for the way the average value of the new group is calculated. Each time an object is added to the group, the new average is calculated as the sum of the similarity measure of the new object and the

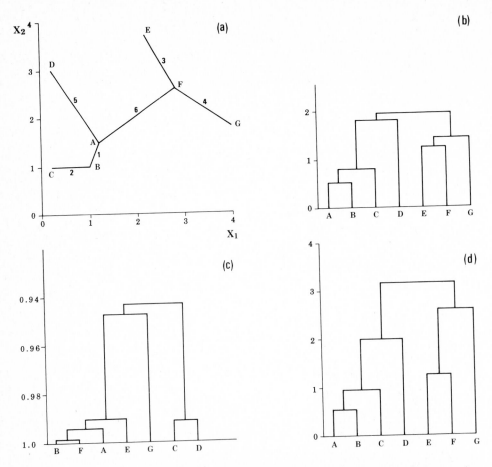

Fig. 10.1. (a) Illustration of the single-linkage method of cluster analysis using Euclidean distances and the hypothetical 2-dimensional data given in Table 10.2. The order in which the objects are joined into a single group is shown by the numbers on the lines. (b) The dendrogram obtained using the single-linkage method and the Euclidean distance. The horizontal scale is arbitrary, but the vertical scale shows the similarity values at which the objects were joined. (c) The dendrogram obtained using the single-linkage method and the cosine theta coefficient. Note the considerable differences in groupings compared to (b), due to the fact that the Euclidean distance and cosine theta coefficient are recording different attributes. (d) The dendrogram obtained using the weighted-pair group method and Euclidean distances. Note that there is no difference in grouping between this and (b), only a difference in the order of linkages and their levels.

previous group average, divided by two. In other words, once objects are grouped together they are thereafter treated as if they were a single object. This means that the group average is weighted in favour of the new object. An example of this method, starting with the similarity matrix of Euclidean distances in Table 10.2, is given in Table 10.3. The 1st linkage is between A and B and the new group average for group AB is given by taking

TABLE 10.3

Series of similarity matrices based on the Euclidean distance to illustrate the weighted-pair group method of linkage, starting with the appropriate matrix in Table 10.2. Each time two objects are linked, the values in the two rows and columns containing the objects are simply averaged to form a single row and column. Although the complete matrix has been rewritten after each linkage for clarity, a computer algorithm would not work in this way. Similarly, most computer programs would only store half of the matrix as it is symmetrical.

Similarity matrix after 1st linkage

	AB	C	D	E	F	G
AB	0.00	0.96	1.98	2.69	2.17	2.96
C	0.96	0.00	2.00	3.36	3.05	3.88
D	1.98	2.00	0.00	2.12	2.63	3.98
E	2.69	3.36	2.12	0.00	1.25	2.62
F	2.17	3.05	2.63	1.25	0.00	1.44
G	2.96	3.88	3.98	2.62	1.44	0.00

Similarity matrix after 2nd linkage

	ABC	D	E	F	G
ABC	0.00	1.99	3.02	2.61	3.42
D	1.99	0.00	2.12	2.63	3.98
E	3.02	2.12	0.00	1.25	2.62
F	2.61	2.63	1.25	0.00	1.44
G	3.42	3.98	2.62	1.44	0.00

Similarity matrix after 3rd linkage

	ABC	D	EF	G
ABC	0.00	1.99	2.82	3.42
D	1.99	0.00	2.37	3.98
EF	2.82	2.37	0.00	2.03
G	3.42	3.98	2.03	0.00

Similarity matrix after 4th linkage

	ABCD	EF	G
ABCD	0.00	2.60	3.70
EF	2.60	0.00	2.03
G	3.70	2.03	0.00

Similarity matrix after 5th linkage

	ABCD	EFG
ABCD	0.00	3.15
EFG	3.15	0.00

the means of the rows and columns of A and B in the original matrix, which gives the 1st reduced matrix in Table 10.3. For example, the new similarity measure between AB & D is given by $(1.80 + 2.15)/2 = 1.98$. The 2nd linkage is between C and AB (0.96), the 3rd between E and F (1.25), the 4th between D and ABC (1.99), the 5th between G and EF (2.03), and the two groups ABCD and EFG are finally linked at a value of 3.15. These linkages are shown in the dendrogram in Fig. 10.1d in which the grouping is very similar to that obtained with the single-linkage method in Fig. 10.1b. The only minor differences are that the order of linking D to ABC and G to EF is interchanged, and the linkage scale is expanded. A worked numerical example of this method is also given by Davis (1973).

As with similarity measures, the choice of the method of linkage is purely subjective, but the most commonly used method in the geological literature appears to be the weighted-pair group average. Gnanadesikan (1977) gives a detailed discussion of some other methods that can be used, together with some well illustrated examples.

As pointed out by Marriott (1974), one of the properties of most agglomerative methods of linkage is that once a link is formed it is never broken. This could prevent two distinct clusters, which are joined by a chain, from being cleanly separated if some of the objects in the chain are very similar and are, therefore, linked at an early stage. Such a situation could arise with the hypothetical data shown in Fig. 10.2, which is based on a diagram by Marriott (1974, p.65). He further points out that although such a distribution of data is unlikely in 2-dimensional space, it is much more likely in higher dimensional space as the number of variables increases.

10.3 CLASSIFICATION PROCEDURE

In principle the method is simple, as all that is required is a measure of similarity between the unknown object to be classified and a group of objects known to belong to the class. In cluster analysis this was achieved by

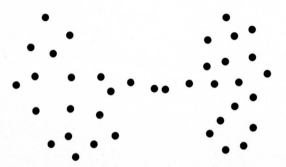

Fig. 10.2. Diagram based on one by Marriott (1974, p.65) to illustrate how two clusters joined by a chain may not be cleanly separated, if objects in the chain are very similar and are joined together at an early stage.

calculating a similarity measure between the unknown and the group mean. However, if correlations exist between the variables this is not entirely satisfactory, as can be seen in Fig. 10.3. If simple Euclidean distance is used, then points Q and R would be judged as being equally similar to the group, whose probability ellipses are shown, as they are equidistant from the group mean, M. However, inspection of the probability ellipses, which are the same as those of Fig. 4.1, indicates that R is much more likely to belong to the group than Q. Similarly, although Q is closer to the group mean than P, they both have equal probabilities of belonging to the group, as they both lie on the same ellipse.

What is required, therefore, is a method of evaluating which probability ellipse, or hyperellipsoid in p-dimensional space, an unknown point lies on. This is relatively simple to do and is based on the theory of the *multivariate normal distribution*. It can be shown (e.g. Tatsuoka, 1971, p.66ff) that the family of probability hyperellipsoids is given by the matrix equation:-

$$\chi^2 = (\mathbf{x} - \bar{\mathbf{x}})^* \, \mathbf{D}^{-1} \, (\mathbf{x} - \bar{\mathbf{x}}) \tag{10.4}$$

where $\mathbf{x}$ and $\bar{\mathbf{x}}$ are vectors of the values of the unknown and group sample mean, respectively, and $\mathbf{D}^{-1}$ is the inverse of the group sample variance-covariance matrix (equation 7.1). The value χ^2, is distributed as an accumulative chi-square with p d.f. and is often referred to as a *generalized distance*. The smaller the value of χ^2, the more likely the unknown is to belong to the group. Table A5 in the Appendix, or any of the standard computer subroutines, can then be used to convert the χ^2 into a probability that the

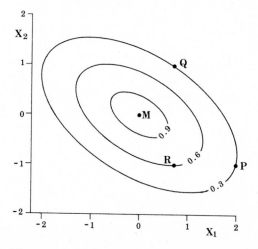

Fig. 10.3. Diagram to illustrate some of the principles of classification procedure, using the same bivariate distribution as Fig. 4.1. The ellipses of equal probability are labelled with the probability of belonging to the population or group. Although Q is geometrically much nearer to the group centroid M than P, they both have a probability of 0.3 of belonging to the group, as they both lie on the same probability ellipse. R, on the other hand, has a higher probability (0.6) of belonging to the group than Q (0.3), even though they are the same geometric distance from the group centroid.

unknown belongs to the group. This step, of course, is only valid if the population is multivariate normally distributed. From what little work has been done on petrological data, Le Maitre (1976c) has shown that for basalts, andesites, dacites and rhyolites the distributions are not multivariate normal, as there are fewer observations in the centre of the distribution and more at the extremities than would be expected. This is in accord with the concepts of rock classifications, where the boundaries have been arbitrarily placed throughout a reasonably continuous distribution, so that there is no reason to suppose that the centre of a "pigeon-hole" should have a higher density of data than elsewhere. However, it is interesting to note that as the empirical distribution curves for these 4 groups (Le Maitre, 1976c) are very similar, assignment rules using the χ^2 can still be used.

Numerical example

To illustrate the method we will use the bivariate distribution in Fig. 10.3, which is the same as that in Fig. 4.1. The standard deviations of x_1 and x_2 are 1.3 and 1.0, respectively, and the correlation coefficient is -0.5. The variance-covariance matrix is, therefore:-

$$\mathbf{D} = \begin{bmatrix} 1.69 & -0.65 \\ -0.65 & 1.00 \end{bmatrix} \quad \text{which gives } \mathbf{D}^{-1} = \begin{bmatrix} 0.7890 & 0.5128 \\ 0.5128 & 1.3333 \end{bmatrix}$$

For simplicity, we will put the origin at the mean of the distribution so that $\bar{\mathbf{x}}$ is a null vector and can be ignored. Let us then evaluate equation 10.4 for P, shown in Fig. 10.3, which has $x_1 = 2.0$ and $x_2 = -1.0$. We can then write:-

$$\chi^2 = \begin{bmatrix} 2.0 & -1.0 \end{bmatrix} \cdot \begin{bmatrix} 0.7890 & 0.5128 \\ 0.5128 & 1.3333 \end{bmatrix} \cdot \begin{bmatrix} 2.0 \\ -1.0 \end{bmatrix}$$

$$= \begin{bmatrix} 1.0652 & -0.3077 \end{bmatrix} \cdot \begin{bmatrix} 2.0 \\ -1.0 \end{bmatrix} = 2.44$$

Similarly, Q with $x_1 = 0.7$ and $x_2 = 1.0$ gives a χ^2 value of 2.44, which means that P and Q lie on the same ellipse and, therefore, have the same probability of belonging to the group. This probability is given by the level of significance corresponding to $\chi^2 = 2.44$ with 2 d.f. in Table A5, which is approximately 0.3. For R, with $x_1 = 0.7$ and $x_2 = -1.0$, we have $\chi^2 = 1.00$ which, using Table A5 again, gives a probability of approximately 0.6. This means that R is twice as likely to belong to the group as Q, even though they are geometrically the same distance from the group mean M.

10.3.1 Application to more than one group

In practice, classification procedure is used to assign unknowns to one of two or more groups. To do this, equation 10.4 has to be calculated for each group, so that it is convenient to generalize the equation by writing it as:-

$$\chi_i^2 = (\mathbf{x} - \bar{\mathbf{x}}_i)^* \, \mathbf{D}_i^{-1} \, (\mathbf{x} - \bar{\mathbf{x}}_i) \qquad (10.5)$$

where the subscripts i refer to the ith group. When all the variance-covariance matrices are the same, an unknown can then be assigned to the group for which χ_i^2 is the smallest. This can be formalized into the following assignment rule:-

$$H_j \text{ if } \quad \chi_j^2 < \chi_i^2 \quad \text{ for } i = 1 \text{ to g}, i \neq j \tag{10.6}$$

where H_j stands for "accept the hypothesis that the unknown belongs to group j", and g is the number of groups. If the χ_i^2 values are converted to probabilities $Pr(\chi_i^2)$, then the assignment rule becomes:-

$$H_j \text{ if } \quad Pr(\chi_j^2) > Pr(\chi_i^2) \quad \text{ for } i = 1 \text{ to g}, i \neq j \tag{10.7}$$

Both of these rules will, of course, give the same answers and will also assign any unknown to one of the groups. However, with rule 10.7 it is possible to impose an additional constraint, that the probability must be greater than a particular value before the hypothesis is accepted. This may be written as:-

$$H_j \text{ if } \quad Pr(\chi_j^2) > Pr(\chi_i^2) \quad \text{ for } i = 1 \text{ to g}, i \neq j, \text{ and } \quad Pr(\chi_j^2) > c \tag{10.8}$$

where c is the specified cut-off probability. Which of these two rules is used depends upon the application. For example, if it is essential that every unknown is assigned to a group, then rule 10.6 or 10.7 should be used. If, however, it is important to know that an unknown is unlikely to belong to any of the groups, as is usually the case in petrology, where most of the classifications are mutually exclusive, then the assignment rule 10.8 could be used. This would be the case, for example, if a series of rock analyses were being classified as rhyolites, trachytes or phonolites, as some of the analyses may belong to none of the 3 groups. It is, therefore, important to understand what type of assignment rule is being used, if using an unfamiliar computer program. Furthermore, some procedures sum the probabilities to unity (or 100%) before applying the assignment rule, so that rule 10.8 would then be misleading. For example, consider the case of a basalt analysis accidentally being included in a series of analyses to be classified as rhyolite, trachyte or phonolite. The raw probabilities may well be 10^{-6}, 10^{-4} and 10^{-6}, respectively. Obviously, the basalt analysis should not be classified as any of the 3 groups, but if the probabilities are recalculated to 100%, the basalt would be classified as a trachyte with a 98% probability! Computer programs written for the social sciences often work this way.

If the variance-covariance matrices are different for each group, as is usually the case, a different assignment rule should be used. This takes into account the fact that the same probability ellipses in different groups, will be of different sizes in space. Under such circumstances, additional terms are added to the assignment rule 10.6 which then becomes:-

$$H_j \text{ if } \quad \chi_j^2 + \log_e|D_j| < \chi_i^2 + \log_e|D_i| \quad \text{ for } i = 1 \text{ to g}, i \neq j \tag{10.9}$$

where $|D_j|$ is the determinant of the matrix $\mathbf{D_j}$. Such a rule, however, does not take into account the fact that an unknown may not belong to any of the groups. In practice, as long as the determinants of the variance-covariance matrices are similar, the additional logarithmic terms will almost cancel out, so that the previous assignment rules can be used. This rule breaks down for closed data, of course, as the determinant of a singular matrix is zero.

More information on these topics can be found in Cooley & Lohnes (1971), Das Gupta (1973), Gnanadesikan (1977) and Tatsuoka (1971). Le Maitre (1976c) uses this approach to present a new type of classification of igneous rocks based on the probability of predefined group membership.

10.3.2 *Using prior probabilities*

The theory of classification procedure also allows the assignment rules to take into account prior probability, which is the probability of obtaining a member of a particular group from a mixed population of g groups. If Pr_i is the prior probability of obtaining a member of the ith group, the assignment rule of 10.9 is further modified and becomes:-

$$H_j \text{ if } \quad \chi_j^2 + \log_e|D_j| - 2\log_e Pr_j \quad < \quad \chi_i^2 + \log_e|D_i| - 2\log_e Pr_i$$
$$\text{for } i = 1 \text{ to } g, \, i \neq j \qquad (10.10)$$

However, this method is open to the criticism that it tends to keep large groups large and small groups small. Furthermore, as the additional terms $2\log_e Pr_j$ can be quite different in magnitude and can, therefore, have a considerable effect on the assignment, it is essential that the prior probabilities are accurately known. In most cases this is unlikely, as the determinations of prior probabilities is usually largely subjective. In view of this it is suggested that prior probabilities should be used with the utmost caution with petrological data as, for example, a basic rock should be classified as a basalt, basanite, etc. on its own characteristics, and not according to the frequency of occurrence of the rock types.

10.3.3 *Reducing the dimensionality of the problem*

With p variables and g groups, the calculation of the χ_i^2 for one unknown, involves up to $gp(p + 1)$ multiplications and additions. If this computation has to be performed a large number of times, it is obviously an advantage to be able to reduce the number of variables being used, and hence the amount of computation. As Tatsuoka (1971, p.232) points out, this can be done by transforming the original p variables into the g - 1 discriminant coordinates, calculated from the discriminant functions obtained from a multiple discriminant analysis of the g groups (see Chapter 9 and equation 9.5). If the variance-covariance matrix for each group is calculated using the discriminant coordinates as variables, the results obtained using the minimum chi-square assignment rule (10.6) will be identical to those that would have been obtained using the original p variables. Theoretically, if the variance-covariance matrices of each group are different, the two sets of variables will give different results. However, the method appears to be fairly robust as, to quote Tatsuoka (1971, p.233) "the two spaces yield closely similar results so long as the dispersion matrices (variance-covariance matrices) are not drastically different".

A further reduction can be achieved by using only the statistically significant discriminant functions in calculating the discriminant coordinates. In this instance, the assignments will differ slightly from those obtained with the p variables, but it has the advantage that chance apparent differences, due to sampling errors, will probably be eliminated.

10.3.4 *The problem of singular matrices*

If the p variables used in the classification procedure are a set of closed data (section 4.2.3), the variance-covariance matrix $\mathbf{D}_i$, will be singular, so that its regular inverse $\mathbf{D}^{-1}$ will not exist and its determinant will be zero. This means that the χ_i^2 of equation 10.5 will not be able to be calculated in the usual way. There are several ways in which this problem can be overcome.

One method is to omit one of the variables and to perform the calculation with p - 1 variables. Fortunately, it does not matter which variable is left out, as the p possible values of χ_i^2 derived from equation 10.5 are all identical. If the χ_i^2 are to be converted to probabilities, remember that the d.f. will be p - 1, i.e. it is still the actual number of variables used. This was the solution adopted by Le Maitre (1976c) for a classification based on the closed major element data of groups of basalts, andesites, dacites and rhyolites. Another method is to transform the variables to discriminant coordinates, as described in the previous section.

The final method to be mentioned is mathematically more elegant and involves using a *generalized inverse* of the variance-covariance matrix, rather than the regular inverse. If $\mathbf{D}$ is a square non-singular matrix, its regular inverse $\mathbf{D}^{-1}$ is defined by the matrix equation:-

$$\mathbf{D}\mathbf{D}^{-1} = \mathbf{D}^{-1}\mathbf{D} = \mathbf{I} \qquad (10.11)$$

where $\mathbf{I}$ is an identity matrix, which has 1 in all the diagonal terms, and 0 in all the off-diagonal terms. However, if $\mathbf{D}$ is a singular matrix, $\mathbf{D}^{-1}$ does not exist, but a generalized inverse $\mathbf{D}^-$ does exist. This is defined by the matrix equation:-

$$\mathbf{D}\mathbf{D}^-\mathbf{D} = \mathbf{D} \qquad (10.12)$$

In fact, given $\mathbf{D}$, which does not have to be a square matrix, there are an infinite number of generalized inverse matrices $\mathbf{D}^-$ which satisfy equation 10.12. However, when $\mathbf{D}$ is a square non-singular matrix, only one generalized inverse exists, and then $\mathbf{D}^{-1} = \mathbf{D}^-$. As Tatsuoka (1971, p.152) points out, in many statistical applications we do not require the full implications of equation 10.11, but can make do with equation 10.12. In otherwords, when $\mathbf{D}^{-1}$ does not exist, it can be replaced by $\mathbf{D}^-$ in equation 10.5. However, if the χ_i^2 value is to be converted to a probability, the d.f. are now p - 1, where p is the number of variables or the order of the generalized inverse matrix $\mathbf{D}^-$.

Although there are many ways of calculating a generalized inverse (e.g. Bjerhammar, 1973; Tatsuoka, 1971), for square matrices a simple method is avaliable (Rao & Mitra, 1971; Williams, 1980). This involves adding a constant, k/p, to each term of the matrix, where k is any value, such as the sum of the diagonal terms of the matrix, and p is the order of the matrix. The resulting non-singular matrix is then inverted by any of the usual methods. The value 1/kp is then subtracted from every term of the regular inverse, to obtain a generalized inverse that is also symmetrical. If this particular generalized inverse matrix is used in equation 10.5, the value of χ_i^2 will be exactly the same as that obtained if any one of the variables was omitted. This is the method used by subroutine GMATIN in the CLAIR data system (section 12.3.1).

Numerical example

The variance-covariance matrix of the previous numerical example can be made singular by adding an extra row and column, such that the row and column sums are zero (equation 7.8), which gives:-

$$\mathbf{D} = \begin{bmatrix} 1.69 & -0.65 & -1.04 \\ -0.65 & 1.00 & -0.35 \\ -1.04 & -0.35 & 1.39 \end{bmatrix}$$

To form the generalized inverse we first add k/p to every term in the matrix. Letting k equal the sum of the diagonal terms we, therefore, add 4.08/3 = 1.36 to each term and obtain:-

$$\mathbf{E} = \begin{bmatrix} 3.05 & 0.71 & 0.32 \\ 0.71 & 2.36 & 1.01 \\ 0.32 & 1.01 & 2.75 \end{bmatrix}$$

This non-singular matrix is then inverted to give the regular inverse as:-

$$\mathbf{E}^{-1} = \begin{bmatrix} 0.3526 & -0.1050 & -0.0025 \\ -0.1050 & 0.5340 & -0.1839 \\ -0.0025 & -0.1839 & 0.4315 \end{bmatrix}$$

Then by subtracting 1/kp = 1/(4.08 x 3) = 0.0817 from each term we obtain the generalized inverse matrix as:-

$$\mathbf{D}^- = \begin{bmatrix} 0.2709 & -0.1867 & -0.0842 \\ -0.1867 & 0.4523 & -0.2656 \\ -0.0842 & -0.2656 & 0.3498 \end{bmatrix}$$

We will now evaluate equation 10.4, using $\mathbf{D}^-$ instead of $\mathbf{D}^{-1}$, which in this case does not exist. With $\bar{\mathbf{x}}$ a null vector as before, the original variables must sum to zero to be closed, so that we must use $x_1 = 2.0$, $x_2 = -1.0$ and $x_3 = -1.0$, which gives:-

$$\chi^2 = \begin{bmatrix} 2.0 & -1.0 & -1.0 \end{bmatrix} \cdot \begin{bmatrix} 0.2709 & -0.1867 & -0.0842 \\ -0.1867 & 0.4523 & -0.2656 \\ -0.0842 & -0.2656 & 0.3498 \end{bmatrix} \cdot \begin{bmatrix} 2.0 \\ -1.0 \\ -1.0 \end{bmatrix}$$

$$= \begin{bmatrix} 0.8127 & -0.5601 & -0.2526 \end{bmatrix} \cdot \begin{bmatrix} 2.0 \\ -1.0 \\ -1.0 \end{bmatrix} = 2.44$$

which is the same as the value obtained for P, in Fig. 10.3, in the previous numerical example. If we omit the variable x_1 from the calculation we have:-

$$\mathbf{D} = \begin{bmatrix} -1.00 & -0.35 \\ -0.35 & 1.39 \end{bmatrix} \qquad \text{which gives } \mathbf{D}^{-1} = \begin{bmatrix} 1.0966 & 0.2761 \\ 0.2761 & 0.7890 \end{bmatrix}$$

Using equation 10.4 with $x_2 = -1.0$ and $x_3 = -1.0$ we then obtain:-

$$\chi^2 = \begin{bmatrix} -1.0 & -1.0 \end{bmatrix} \cdot \begin{bmatrix} 1.0966 & 0.2761 \\ 0.2761 & 0.7890 \end{bmatrix} \cdot \begin{bmatrix} -1.0 \\ -1.0 \end{bmatrix}$$

$$= \begin{bmatrix} -1.3727 & -1.0651 \end{bmatrix} \cdot \begin{bmatrix} -1.0 \\ -1.0 \end{bmatrix} = 2.44$$

which again is the same value as before. This illustrates that it does not matter which variable is omitted from the calculation, if the data are closed.

CHAPTER 11

PROPAGATION OF ERRORS

11.1 INTRODUCTION

In previous chapters, standard deviations have been calculated from samples of n observations, using an equation such as 2.2. In most instances, the variables were primary, but in this chapter we are interested in the standard deviations of secondary variables when we know the standard deviations of the primary variables, x_1, x_2, ... x_p, from which they are calculated. This process, whereby errors in primary variables are transmitted to secondary variables, is known as the *propagation of errors*. To simplify the notation, all the relevant formulae are expressed in terms of variance rather than standard deviation.

In mathematical terms, if the secondary variable, w, is a function of x_1, x_2, ... x_p, which can be written as:-

$$w = f(x_1, x_2, ... x_p) \tag{11.1}$$

then the variance of w can be estimated using the approximation (e.g. Kendall, 1943, p.208) that:-

$$\text{var } w = \sum \left[\frac{\partial w}{\partial x_i}\right]^2 \text{var } x_i + \sum \left[\frac{\partial w}{\partial x_i} \frac{\partial w}{\partial x_j}\right] \text{cov } x_i x_j \tag{11.2}$$

where var x_i is the variance of the ith variable, and cov $x_i x_j$ is the covariance between the ith and jth variable. The 1st summation is taken over all p values of i, and the 2nd summation is over all p values of i and j such that $i \neq j$. If it is assumed that the covariances are zero, i.e. the variables are independent, then equation 11.2 can be simplified to:-

$$\text{var } w = \sum \left[\frac{\partial w}{\partial x_i}\right]^2 \text{var } x_i \tag{11.3}$$

which is the expression that is commonly used for calculating the variance of a function. However, with closed data of the constant sum type, we know from equation 4.12 that not all the covariances can be zero, and that their sum must be negative. This means that if the covariance terms in equation 11.2 are ignored, the absence of their overall negative contribution will tend to give an overestimate of the variance. As this is generally more acceptable than an underestimate of the variance, the use of equation 11.3 for closed data would, therefore, seem reasonable.

Before proceeding further, it is useful to consider a few simple results that follow directly from equation 11.3. They are:-

1. If $w = ax_1$, where a is a constant, then:-

$$\text{var } w = a^2 \text{ var } x_1 \tag{11.4}$$

2. If $w = x_1 + x_2$ or $w = x_1 - x_2$, then:-

$$\text{var } w = \text{var } x_1 + \text{var } x_2 \tag{11.5}$$

3. If $w = x_1 x_2$ or $w = x_1 / x_2$, then:-

$$\frac{\text{var } w}{w^2} = \frac{\text{var } x_1}{x_1^2} + \frac{\text{var } x_2}{x_2^2} \tag{11.6}$$

4. If $w = (x_1 + x_2) / (x_3 + x_4)$, then:-

$$\frac{\text{var } w}{w^2} = \frac{(\text{var } x_1 + \text{var } x_2)}{(x_1 + x_2)^2} + \frac{(\text{var } x_3 + \text{var } x_4)}{(x_3 + x_4)^2} \tag{11.7}$$

11.2 COUNTING X-RAYS AND OTHER RADIATIONS

As many methods of chemical analysis now involve counting x-rays, or other types of radiation, it is important to understand the errors involved which can then be used to judge the precision of the measurements. The generation of x-rays and other types of radiation follows the Poisson distribution, which has the property that its variance is equal to its mean (section 1.3.4). Furthermore, as the number of counts measured is usually large, the normal distribution can be used as an approximation to the Poisson distribution.

11.2.1 Single counts

Suppose that in a given time T, C counts have been recorded. If C is then used as an estimate of the mean of the population, C will also be an estimate of the variance, so that we can state the $100(1 - a)\%$ confidence limits as:-

$$C \pm z_{a/2} \sqrt{C} \tag{11.8}$$

where $z_{a/2}$ is the value of z, from Table A2 in the Appendix, which corresponds to an area of $a/2$. Alternatively, this value may be obtained from the bottom row of Table A3. For example, for the 95% and 99% confidence limits the values of $z_{a/2}$ would be 1.960 and 2.576, respectively.

Based on equation 2.5, the *percentage relative error* of the measurement is then given by:-

$$E = 100 \sqrt{C} / C = 100 / \sqrt{C} \tag{11.9}$$

For convenience, the upper and lower 99% confidence limits and the percentage relative errors for a range of counts are given in Table 11.1.

TABLE 11.1

The upper and lower 99% confidence limits for a range of x-ray counts calculated using equation 11.8, together with their percentage relative errors as defined by equation 11.9.

Lower limit	Number of counts	Upper limit	Percentage relative error
74	100	126	10.00
442	500	558	4.472
919	1,000	1,081	3.162
4,818	5,000	5,182	1.414
9,742	10,000	10,258	1.000
49,424	50,000	50,576	0.447
99,185	100,000	100,815	0.316
498,178	500,000	501,822	0.141
997,424	1,000,000	1,002,576	0.100
4,994,240	5,000,000	5,005,760	0.045
9,991,854	10,000,000	10,008,146	0.032

Suppose we now record n separate counts C_1, C_2, .. C_n, each for time T. The mean number of counts $\bar{C}$ for time T is then:-

$$\bar{C} = \sum C_i / n \tag{11.10}$$

However, the variance of $\sum C_i$ is $\sum C_i$, so that using equation 11.4 we obtain the variance of $\bar{C}$ as $\sum C_i / n^2$, which is equal to $\bar{C}/n$. This gives a percentage relative error for $\bar{C}$ of:-

$$E = 100 \sqrt{(\bar{C} / n)} / \bar{C} = 100 / \sqrt{(n\bar{C})} \tag{11.11}$$

This value, however, is also the percentage relative error that would have been obtained if we had recorded $n\bar{C}$ counts. Hence from the point of view of the percentage relative error, it is immaterial whether we count n separate times for time T, or once for time nT. From a practical point of view, however, there is an advantage in counting n times, as a check can then be made on the stability of the experimental conditions, as shown in the next section.

In most cases we are not really interested in the actual number of counts, but are concerned with the count rate, R, where:-

$$R = C / T \tag{11.12}$$

If we then assume that the time T can be measured without error and, therefore, can be regarded as a constant, we can use equation 11.4 to give the variance of R as:-

$$\text{var } R = C / T^2 = RT / T^2 = R / T \tag{11.13}$$

This means that we can reduce the standard deviation of R to any value we like by a suitable choice of counting time T. Note, that to reduce the standard deviation of R by 2, we must increase the counting time by 4.

Numerical example

If 1600 counts are recorded in 1 minute, the 99% confidence limits are given by equation 11.8 as:-

$1600 \pm 2.576 \times \sqrt{1600} = 1600 \pm 103 = 1497$ and 1703

The percentage relative error = 100/40 = 2.5%. We can also use equation 11.9 to obtain the number of counts we must record to achieve a particular percentage relative error. For example, to obtain a percentage relative error of 1% we must record a minimum of $(100/1)^2 = 10000$ counts, which would mean counting for over 6 minutes.

11.2.2 *Testing count stability*

The confidence limits of equation 11.8 can also be used to test the stability of the counting equipment and the source of radiation, by repeatedly counting for a fixed time T. If more than $100(1 - \alpha)\%$ of the counts recorded are within the limits of equation 11.8, we can conclude that there is no evidence of instability in the system. In this instance $\bar{C}$ would be the value used in equation 11.8.

Numerical example

To illustrate the method of testing the stability of counts, we will use the data of Table 11.2. The mean of the counts is 402, so that we may use equation 11.8, to obtain the 99% confidence limits as:-

$402 \pm 2.576 \sqrt{402} = 402 \pm 52$ or   350 and 452

As all of the counts are within these limits, we can then conclude that there is no evidence to suggest that there is any instability in the count rate.

TABLE 11.2

Some data to illustrate how the stability of x-ray counting equipment may be tested. The counts were recorded from a standard in the electron microprobe.

Number of counts in 10 seconds				
405	392	435	406	401
413	378	418	359	422
375	389	447	396	394
422	409	410	391	377

Total number of counts = 8039
Sample mean = 8039/20 = 402

11.2.3 *Optimizing peak and background counting times*

In accurate analytical work, background corrections have to be made before x_u, the concentration of an element or oxide in an unknown, can be calculated. This usually involves using a formula of the type:-

$$x_u = (R_p - R_b) / K \tag{11.14}$$

where R_p and R_b are the count rates on the peak and background, respectively, and K is the calibration constant for the particular set of experimental conditions, expressed as count rate per concentration of element or oxide. As will be seen later, K is also subject to error. Now in measuring the peak and background count rates, there is no reason why they should both be counted for the same length of time, although many operating systems do. If a fixed amount of time, T, is available for counting both the peak and background, there is an optimum division of time between the peak, T_p, and background, T_b, to minimize the errors in the corrected peak count rate, R_c, where:-

$$R_c = R_p - R_b \tag{11.15}$$

Using equations 11.13 and 11.5, we can then write the variance of R_c as:-

$$\text{var } R_c = R_p/T_p + R_b/T_b \tag{11.16}$$

which, as $T = T_p + T_b$, can be written as:-

$$\text{var } R_c = R_p/T_p + R_b/(T - T_p) \tag{11.17}$$

Differentiating this with respect to T_p, equating the result to zero, and rearranging the equation then gives:-

$$T_p/T_b = \sqrt{(R_p/R_b)} \tag{11.18}$$

as the relationship that will minimize the variance of R_c. This can be expressed in a more convenient form as:-

$$T_p = T \sqrt{R_p} / (\sqrt{R_p} + \sqrt{R_b}) \tag{11.19}$$

so that given T, the total time available, T_p, the optimum time for counting the peak, can be determined. In practice such a strategy is only worth using if the system is under computer control, so that approximate count rates can be determined prior to the main counting episode. However, the equations do show that it is largely a waste of time counting the background for as long as the peak, unless the two are almost equal, i.e. one is working near the lower limit of detection. Table 11.3 tabulates the peak to background count rate ratios that correspond to particular percentage amounts of total time to be spent counting the peak for optimum conditions.

TABLE 11.3

This table can be used to optimize the counting times between peak and background, or between standard and unknown. Using the symbols at the top of the table, it can be used to determine T_p, the percentage of total time that must be spent counting a peak in order to minimize the variance of the corrected peak height count rate, given the ratio of the peak to background count rates, R_p/R_b. This is based on equation 11.18. Using the symbols at the bottom of the table, it can be used to determine T_u, the percentage of total time that must be spent counting an unknown in order to minimize the variance of the nominal concentration of an element or oxide in the unknown, given the ratio of the standard and unknown count rates R_s/R_u. This is based on equation 11.28.

Key to optimize PEAK TO BACKGROUND counting time

R_p/R_b	T_p	R_p/R_b	T_p	R_p/R_b	T_p	R_p/R_b	T_p	R_p/R_b	T_p
1.00	50	2.25	60	5.44	70	16.0	80	81.	90
1.17	52	2.66	62	6.61	72	20.8	82	132.	92
1.38	54	3.16	64	8.10	74	27.6	84	245.	94
1.62	56	3.77	66	10.0	76	37.7	86	576.	96
1.91	58	4.52	68	12.6	78	53.8	88	2401.	98
R_s/R_u	T_u	R_s/R_u	T_u	R_s/R_u	T_u	R_s/R_u	T_u	R_s/R_u	T_u

Key to optimize STANDARD TO UNKNOWN counting time

Numerical example

An electron microprobe, records 16275 counts from the Si peak on a SiO_2 standard and 1320 counts from the background. The counting time in both instances is 60 seconds. What is the standard deviation of the corrected peak height count rate R_c, expressed as counts per second? Using equation 11.12 we have:-

$R_p = 16275/60 = 271.25$ and $R_b = 1320/60 = 22.0$

so that using equation 11.16 we obtain:-

var $R_c = 271.25/60 + 22.0/60 = 4.89$

which gives the standard deviation as 2.21 counts per second. As the total counting time was 120 seconds, what is the optimum partition of time between the peak and background to minimize the standard deviation of R_c? Using equation 11.19 we have:-

$T_p = 120 \sqrt{271.25} / (\sqrt{271.25} + \sqrt{22.0}) = 93$

which gives $T_b = 120 - 93 = 27$ seconds. The reader may care to check this with Table 11.3, where a ratio of $R_p/R_b = 12.3$ gives a percentage time of 78% for counting the peak, or 0.78 x 120 = 93 seconds. Again using equation 11.16, we can then obtain the minimum standard deviation of R_c for a total counting time of 120 seconds as:-

$$\text{var } R_c = 271.25/93 + 22.0/27 = 3.73$$

or a standard deviation of 1.93 counts per second. Whether or not the improvement is worth the extra effort, is a question only the investigator can decide. However, the example does show that the optimum times are not very critical.

11.2.4 Measuring concentration in an unknown

The calibration constant, K, used in equation 11.14 is also subject to error and in many techniques, such as electron microprobe and x-ray fluorescence analysis, is initially determined from measurements on a single standard by an equation of the type:-

$$K = (R_{ps} - R_{bs}) / x_s \tag{11.20}$$

where R_{ps} and R_{bs} are the count rates on the standard peak and background, respectively, and x_s is the concentration of the element or oxide in the standard, which is also subject to error (see Table 1.1). Inserting this value of K into equation 11.14, and expanding the notation, we obtain:-

$$x_u = x_s (R_{pu} - R_{bu}) / (R_{ps} - R_{bs}) \tag{11.21}$$

where the subscripts u, s, p and b, refer to the unknown, standard, peak and background, respectively. This is the basic equation used by many analytical techniques which involve counting procedures to determine the nominal concentration of an element or oxide in an unknown. Using equations 11.7 and 11.5, we can then determine the variance of x_u as:-

$$\text{var } x_u = x_u^2 \left[\frac{(R_{pu}/T_{pu} + R_{bu}/T_{bu})}{(R_{pu} - R_{bu})^2} + \frac{(R_{ps}/T_{ps} + R_{bs}/T_{bs})}{(R_{ps} - R_{bs})^2} + \frac{\text{var } x_s}{x_s^2} \right] \tag{11.22}$$

As can be seen, the longer the counting times, the smaller the variance of the nominal concentration in the unknown, x_u, becomes. In general, if all the counting times are increased by a factor of 4, the standard deviation will be reduced by 2, assuming that var x_s is zero. However, in the limiting case, when the counting times are infinitely large, it is clear that the expression var x_u/x_u^2 can never be less than var x_s/x_s^2, i.e. the relative error of x_u can never be less than the relative error of x_s.

If the peak and background counting times are the same for both the standard and unknown, the equation can be multiplied top and bottom by T^2, where T is the common counting time, and rearranged, using equation 11.12, to give:-

$$\text{var } x_u = x_u^2 \left[\frac{C_{pu} + C_{bu}}{(C_{pu} - C_{bu})^2} + \frac{C_{ps} + C_{bs}}{(C_{ps} - C_{bs})^2} + \frac{\text{var } x_s}{x_s^2} \right] \tag{11.23}$$

Note that the two left-most terms in equations 11.22 and 11.23, are due entirely to counting errors and, therefore, effect the precision of the measurement. The factor var x_S/x_S^2, however, contributes to the accuracy of the measurement.

If all the measurements have been of major oxides, it is then possible to use equation 11.5 to calculate the variance of the analysis total, as:-

$$\text{var Total} = \sum \text{var } x_{ui} \tag{11.24}$$

where x_{ui} is the concentration of the ith oxide in the unknown. The standard deviation of the total should be included in all computer output derived from electron microprobe and x-ray fluorescence analysis, as one check on whether or not the analysis is acceptable.

In electron microprobe and x-ray fluorescence analysis the nominal concentrations are further corrected to produce the actual concentrations. This means that the variance of the final results will be slightly different to that given by equation 11.22. However, for all practical purposes this difference can be ignored, unless extremely large corrections are applied.

Numerical example

The following data were obtained from an electron microprobe when determining CaO in a pyroxene, and using a wollastonite standard containing 47.8% CaO.

	Counts	Time seconds	Counts per second
Standard peak	110055	30	3668.5
Standard background	118	9	13.1
Pyroxene peak	47586	30	1586.2
Pyroxene background	90	9	10.0

The nominal concentration of CaO in the pyroxene can then be obtained using equation 11.21 which gives:-

$$\text{CaO} = 47.8 \, (1586.2 - 10.0) / (3668.5 - 13.1) = 20.61$$

Similarly, the standard deviation of CaO can be obtained using equation 11.22 which, assuming var x_S is zero, gives:-

$$\text{var CaO} = 20.61^2 \left[\frac{(1586.2/30 + 10.0/9)}{(1586.2 - 10.0)^2} + \frac{(3668.5/30 + 13.1/9)}{(3668.5 - 13.1)^2} \right]$$

$$= 424.7721 \, (0.000021729 + 0.000009261) = 0.01316$$

which gives the standard deviation of CaO as 0.11. The percentage relative error is then given by 100.0 x $\sqrt{0.01316}$ / 20.61 = 0.56. This calculation is summarized in Table 11.4, with similar data from a series of typical major oxide determinations of 3 minerals and 3 rocks using both electron microprobe and x-ray fluorescence analysis. It should be noted that the x-ray fluorescence

TABLE 11.4

Some examples of the calculation of standard deviations (s_i), using equation 11.22, and percentage relative errors (P.E.) of some major oxides using electron microprobe and x-ray fluorescence analysis. The typical operating condidions under which the measurements were made are also given. Other abbreviations are:- R_p = peak count rate on the standard; R_b = background count rate on the standard; T_p = peak counting time; T_b = background counting time; x_i = nominal concentration.

Typical operating conditions

	Electron microprobe				X-ray fluorescence			
	R_p	T_p	R_b	T_b	R_p	T_p	R_b	T_b
SiO_2	2100	30	31	9	47000	24	120	24
TiO_2	890	30	2	9	12000	24	210	24
Al_2O_3	5500	30	27	9	21000	24	190	24
FeO	1300	30	6	9	43000	24	290	24
MnO	9100	30	72	9	5500	60	290	60
MgO	5000	30	31	9	5000	60	11	60
CaO	3600	30	13	9	30000	24	350	24
Na_2O	500	30	13	9	8200	60	5	60
K_2O	1400	30	15	9	33000	24	170	24

Results for electron microprobe analysis

	Clinopyroxene			Garnet			Fe-Ti-Mn oxide		
	x_i	s_i	P.E.	x_i	s_i	P.E.	x_i	s_i	P.E.
SiO_2	52.28	0.302	0.58	38.78	0.243	0.63	0.10	0.054	54.27
TiO_2	0.34	0.061	17.88	0.28	0.056	19.83	75.72	0.711	0.94
Al_2O_3	5.20	0.066	1.28	21.91	0.131	0.60	0.07	0.033	50.08
FeO	3.90	0.110	2.82	20.44	0.249	1.22	20.02	0.250	1.25
MnO	0.07	0.023	32.67	0.38	0.025	6.61	4.16	0.047	1.14
MgO	14.92	0.111	0.75	10.01	0.091	0.91	0.04	0.032	88.97
CaO	20.61	0.115	0.56	6.27	0.058	0.92	0.04	0.018	42.10
Na_2O	0.83	0.054	6.45	0.07	0.047	65.19	0.01	0.041	417.33
K_2O	0.03	0.013	38.63	0.02	0.015	90.89	0.01	0.017	209.59
Total	98.19	0.375	0.38	98.15	0.395	0.40	100.16	0.760	0.76

Results for x-ray fluorescence analysis

	Dunite			Basalt			Granite		
	x_i	s_i	P.E.	x_i	s_i	P.E.	x_i	s_i	P.E.
SiO_2	34.91	0.051	0.15	51.85	0.068	0.13	70.96	0.087	0.12
TiO_2	0.04	0.001	1.76	1.27	0.004	0.28	0.49	0.002	0.39
Al_2O_3	0.15	0.005	3.22	14.16	0.032	0.23	15.78	0.035	0.22
FeO	18.58	0.022	0.12	8.52	0.012	0.14	2.82	0.005	0.19
MnO	0.27	0.001	0.36	0.15	0.001	0.48	0.04	0.000	1.23
MgO	43.24	0.107	0.25	7.76	0.034	0.44	0.76	0.011	1.40
CaO	0.29	0.002	0.65	9.78	0.015	0.15	1.85	0.005	0.25
Na_2O	0.05	0.008	14.05	2.77	0.028	1.01	4.18	0.034	0.82
K_2O	0.01	0.000	4.90	1.48	0.003	0.20	4.43	0.007	0.15
Total	97.55	0.121	0.12	97.75	0.090	0.09	101.32	0.101	0.10

analyses have smaller errors than the electron microprobe analyses, due to their higher count rates. Several of the electron microprobe determinations are also not significantly different from zero, and it is debatable whether these values should be reported as determined or as zero.

11.2.5 *Optimizing unknown and standard counting times*

Just as the total time can be partitioned between peak and background counts to minimize the error, so it is possible to partition the counting time between an unknown and a standard. If it is assumed that the peak and background counting times have already been partitioned according to equation 11.18, the two numerators involving times in equation 11.22 can be rewritten with simplified subscripts as:-

$$R_p/T_p + R_b/T_b = (\sqrt{R_p} + \sqrt{R_b})^2 / T \tag{11.25}$$

where T is the total time. Equation 11.22 can then be written in terms of T_u and T_s, the total amounts of time spent counting on the unknown and standard, respectively, to give:-

$$\text{var } x_u = x_u^2 \left[\frac{(\sqrt{R_{pu}} + \sqrt{R_{bu}})^2}{T_u(R_{pu} - R_{bu})^2} + \frac{(\sqrt{R_{ps}} + \sqrt{R_{bs}})^2}{T_s(R_{ps} - R_{bs})^2} + \frac{\text{var } x_s}{x_s^2} \right] \tag{11.26}$$

Replacing T_u by the total time available minus T_s, differentiating with respect to T_s and equating the result to zero, gives the relationship that minimizes the variance of x_u as:-

$$\frac{T_s}{T_u} = \frac{(R_{pu} - R_{bu})(\sqrt{R_{ps}} + \sqrt{R_{bs}})}{(\sqrt{R_{pu}} + \sqrt{R_{bu}})(R_{ps} - R_{bs})} \tag{11.27}$$

If it is assumed that the count rates on the two backgrounds are small compared to the peaks, this can be approximated to:-

$$T_s / T_u = \sqrt{(R_{pu} / R_{ps})} \tag{11.28}$$

Note that this gives longer counting time to the lower count rate, which is the opposite of the relationship to optimize the peak to background counting times (equation 11.18), which gives the longer time to the higher count rate. Bearing this in mind, Table 11.3 can also be used to obtain the optimum counting times for the standard and the unknown, if the approximation of equation 11.28 is assumed to be valid.

Numerical example

Using the CaO data of the previous numerical example, we will find the optimum total times to count the standard and the unknown by using equation 11.27, which gives:-

$$\frac{T_s}{T_u} = \frac{(1586.2 - 10.0)(\sqrt{3668.5} + \sqrt{13.1})}{(\sqrt{1586.2} + \sqrt{10.0})(3668.5 - 13.1)} = 0.64$$

This means that $100/(1 + 0.64)$ or 61% of the total time should be spent counting the unknown and 39% counting the standard. However, as the count rates of the two backgrounds are small compared to the peaks, we are justified in using equation 11.28 which gives:-

$$T_s/T_u = \sqrt{(1586.2 / 3668.5)} = 0.66$$

or 60% of the total time spent counting the unknown, which for all practical purposes can be regarded as an identical result. Using this partitioning of the total time of 78 seconds, the standard deviation of CaO is reduced to 0.1046, which is about 10% less than the value of 0.1147 found in the previous example. It must be left to the investigator to decide whether or not this improvement is worthwhile.

11.2.6 Lower limit of detection

For most analytical methods involving counting, the lower limit of detection is governed by the time for which one is prepared to count, which in turn depends upon the long term stability of the equipment being used. The lower limit of detection is found by considering the minimum size of peak that statistically can just be distinguished above the background.

Let the number of counts recorded in time T for the background be C_b, and for the peak be C_p. As we are operating at the lower limit of detection, where the count rates for the peak and background will be very similar, we will assume that equal counting times are used for both the peak and background. The corrected peak height $C_c = C_p - C_b$ will have a variance of $C_p + C_b$ (equation 11.5) so that we can state, at the $100(1 - \alpha)$% confidence level, that if:-

$$C_p - C_b > z_\alpha \sqrt{(C_p + C_b)} \qquad (11.29)$$

the peak height is significantly greater than zero, so that there is evidence to suggest that a peak exists. As this is a one-sided test, and not a two-sided test as with the confidence limits, z_α is the value of z, in Table A2 in the Appendix, for which the area is α. For example, at the 95% and 99% confidence levels the values of z_α are 1.645 and 2.326, respectively.

Now the smallest peak we can be certain of measuring is given by the limiting case of equation 11.29 when:-

$$C_p - C_b = z_\alpha \sqrt{(C_p + C_b)} \qquad (11.30)$$

Squaring both sides and rearranging the terms, we obtain the quadratic equation:-

$$C_p^2 - C_p(2C_b^2 + z_\alpha^2) + C_b(C_b - z_\alpha^2) = 0 \qquad (11.31)$$

Hence, given a background count, C_b, the equation can be solved to find C_p, the smallest peak that can be measured. Although there are two solutions for C_p, only the value greater than C_b is required, which is given by:-

$$C_p = 0.5 \left[2C_b + z_\alpha{}^2 + z_\alpha \sqrt{(8C_b + z_\alpha{}^2)} \right] \tag{11.32}$$

This gives the smallest corrected peak height that can be measured as:-

$$C_c = C_p - C_b = 0.5 z_\alpha \left[z_\alpha + \sqrt{(8C_b + z_\alpha{}^2)} \right] \tag{11.33}$$

Using equation 11.14, the lower limit of detection, x_m, is then given by:-

$$x_m = 0.5 z_\alpha \left[z_\alpha + \sqrt{(8C_b + z_\alpha{}^2)} \right] / KT \tag{11.34}$$

where K is the number of counts per unit time per nominal concentration of the element or oxide. However, as the number of counts recorded on the background is obviously dependent upon the counting time, C_b can be replaced by R_bT to give:-

$$x_m = 0.5 z_\alpha \left[z_\alpha + \sqrt{(8R_bT + z_\alpha{}^2)} \right] / KT \tag{11.35}$$

which gives the relationship between time and the lower limit of detection. However, as z_α is usually small compared with R_b and T, this can be approximated to:-

$$x_m = z_\alpha \sqrt{2R_b} / K \sqrt{T} \tag{11.36}$$

This estimate of the lower limit of detection will always be slightly more optimistic than that given by equation 11.35. Note that this approximation also indicates that the lower limit of detection is inversely proportional to the square root of the counting time.

Using equation 11.36, the lower limits of detection, at the 99% confidence level, have been tabulated in Table 11.5 for a range of values of R_b and K, and for the two specific counting times of 30 and 60 seconds. This will enable rapid estimations to be made of the lower limits of detection for a variety of operating conditions. For example, if K = 300 counts per second per nominal percent and R_b = 10 counts per second, then Table 11.5 gives the lower limit of detection as 0.0045% or 45 ppm for a 60 second counting time. If the counting time were one quarter of this, or 15 seconds, the lower limit of detection would be doubled to 90 ppm.

Numerical example

Again using the CaO data from the numerical example of p.185, we will determine x_m the lower limit of detection for CaO at the 99% confidence level. Using equation 11.20 we have that, K, the calibration constant is given by:-

K = (3668.5 - 13.1) / 47.8 = 76.47

so that using equation 11.35 we obtain:-

$$x_m = 0.5 \times 2.326 \left[2.326 + \sqrt{(8 \times 30 \times 10 + 2.326^2)} \right] / (76.46 \times 30)$$

$$= 0.0260\% \quad \text{or} \quad 260 \text{ ppm}$$

TABLE 11.5

Lower limits of detection, at the 99% confidence level, for a variety of values of background count rate, R_b, and the calibration constant, K, calculated using equation 11.36. The units of R_b must be counts per second, and the units of K counts per second per nominal percent of element or oxide present, to obtain the lower limit of detection as a percentage. Note that 0.0001% is 1 ppm.

Lower limit of detection for 30 second counting time

K	R_b, background count rate						
	1	3	10	30	100	300	1000
1	0.6006	1.0402	1.8992	3.2895	6.0057	10.402	18.992
3	0.2002	0.3467	0.6331	1.0965	2.0019	3.4674	6.3306
10	0.0601	0.1040	0.1899	0.3289	0.6006	1.0402	1.8992
30	0.0200	0.0347	0.0633	0.1096	0.2002	0.3467	0.6331
100	0.0060	0.0104	0.0190	0.0329	0.0601	0.1040	0.1899
300	0.0020	0.0035	0.0063	0.0110	0.0200	0.0347	0.0633
1000	0.0006	0.0010	0.0019	0.0033	0.0060	0.0104	0.0190
3000	0.0002	0.0003	0.0006	0.0011	0.0020	0.0035	0.0063
10000	0.00006	0.0001	0.0002	0.0003	0.0006	0.0010	0.0019

Lower limit of detection for 60 second counting time

K	R_b, background count rate						
	1	3	10	30	100	300	1000
1	0.4247	0.7355	1.3429	2.3260	4.2467	7.3555	13.429
3	0.1416	0.2452	0.4476	0.7753	1.4156	2.4518	4.4764
10	0.0425	0.0736	0.1343	0.2326	0.4247	0.7355	1.3429
30	0.0142	0.0245	0.0448	0.0775	0.1416	0.2452	0.4476
100	0.0042	0.0074	0.0134	0.0233	0.0425	0.0736	0.1343
300	0.0014	0.0025	0.0045	0.0078	0.0142	0.0245	0.0448
1000	0.0004	0.0007	0.0013	0.0023	0.0042	0.0074	0.0134
3000	0.0001	0.0002	0.0004	0.0008	0.0014	0.0025	0.0045
10000	0.00004	0.0001	0.0001	0.0002	0.0004	0.0007	0.0013

Using the approximate formula of equation 11.36 we obtain:-

$$x_m = 2.326 \, \sqrt{(2 \times 10)} / (76.47 \times \sqrt{30}) = 0.0248\% \quad \text{or} \quad 248 \text{ ppm}$$

which for most practical purposes can be considered the same as the previous answer.

To give further indications of the orders of magnitude of the lower limits of detection of some major oxides, as determined by electron microprobe and x-ray fluorescence analysis, Table 11.6 has been compiled, using the typical operating conditions given in Table 11.4.

TABLE 11.6

Some typical lower limits of detection for electron microprobe and x-ray fluorescence analyses of some major oxides, using the 99% confidence level, equation 11.36, and the operating conditions of Table 11.4. K is the counts per second per nominal percent of the oxide, R_b the background count rate in counts per second, T the background counting time in seconds, and L.D. the lower limit of detection as a percentage. Note that 0.0001% is 1 ppm.

Typical lower limits of detection for electron microprobe analysis

	SiO_2	TiO_2	Al_2O_3	FeO	MnO	MgO	CaO	Na_2O	K_2O
K	41	9	55	13	91	50	75	33	66
R_b	31	2	27	6	72	31	13	13	15
T	30	30	30	30	30	30	30	30	30
L.D.	0.0810	0.0954	0.0567	0.1132	0.0560	0.0669	0.0288	0.0654	0.0354

Typical lower limits of dectection for x-ray fluorescence analysis

	SiO_2	TiO_2	Al_2O_3	FeO	MnO	MgO	CaO	Na_2O	K_2O
K	933	8163	935	5741	8661	136	5034	64	10577
R_b	120	210	190	290	290	11	350	5	170
T	24	24	24	24	60	60	24	60	24
L.D.	0.0079	0.0012	0.0099	0.0020	0.0008	0.0104	0.0025	0.0149	0.0008

11.3 MINERAL CALCULATIONS

Although the recalculation of mineral analyses into cations is common practice, little attempt has been made to examine the errors involved in the recalculated values. This section will, therefore, deal with the theory of these errors which is relatively simple. However, as the calculations are somewhat lengthy they are best performed by computer, which is why the equations are given in an easily computable form. The standard deviations of the raw oxides must be known, but if they cannot be calculated by using equations 11.22 or 11.23, they can be assumed.

11.3.1 Based on a fixed number of oxygens

The basic mineral calculation is described in many textbooks (e.g. Deer et al., 1966, p.515) and if done by hand, starts by dividing the weight percent of the oxides by their molecular weights to obtain the molecular proportions. Each molecular proportion is then multiplied by the number of cations and oxygens in its respective oxide, to give the number of cations and oxygens which that oxide contributes to the analysis. However, when the calculation is performed by computer it is much more efficient to perform these two steps in one, by multiplying by a single constant.

If b_i is the number of oxygens in the ith oxide divided by the molecular weight of the ith oxide, the raw sum of oxygens, S_o, is then given by:-

$$S_o = \sum x_i b_i \tag{11.37}$$

where x_i is the weight percent of the ith oxide. Similarly, the raw amount of the ith cation is given by $x_i a_i$, where a_i is the number of cations in the ith oxide divided by the molecular weight of the ith oxide, so that S_c the raw sum of cations is given by:-

$$S_c = \sum x_i a_i \tag{11.38}$$

It then follows that k_i, the number of ith cations calculated on a basis of N_o oxygens is given by:-

$$k_i = N_o x_i a_i / S_o = N_o x_i a_i / \sum x_i b_i \tag{11.39}$$

These steps are tabulated in Table 11.7, using a grossular garnet analysis, modified from an analysis in Deer et al. (1966, p.24, No.3) by the omission of water and the conversion of all the iron to Fe_2O_3, to simulate an electron microprobe analysis.

Now although the variance of k_i can be found from equation 11.39, by differentiation with respect to all the x_i and using equation 11.3, it is easier to

TABLE 11.7

Example of the recalculation of a mineral analysis into cations on a basis of 24 oxygens, using a garnet analysis modified from Deer et al. (1966, p.24, No 3.) and the molecular weights of Table A9 in the Appendix. The standard deviations, s_i, are estimated from the data of Table 11.4, and although they are not used in the mineral recalculation they will be used in later standard deviation calculations. For further details see text.

	x_i	s_i	a_i	b_i		$x_i a_i$	$x_i b_i$	k_i	k_i'
SiO_2	38.69	0.25	0.016643	0.033286	Si	0.6439	1.2878	5.9066	5.9712
TiO_2	0.55	0.06	0.012516	0.025032	Ti	0.0069	0.0138	0.0631	0.0638
Al_2O_3	18.17	0.15	0.019615	0.029423	Al	0.3564	0.5346	3.2693	3.3050
Fe_2O_3	9.90	0.20	0.012524	0.018786	Fe^3	0.1240	0.1860	1.1373	0.6250
FeO	0.00	0.00	0.013919	0.013919	Fe^2	0.0000	0.0000	0.0000	0.5248
MnO	0.64	0.02	0.014097	0.014097	Mn	0.0090	0.0090	0.0828	0.0837
MgO	0.76	0.09	0.024811	0.024811	Mg	0.0189	0.0189	0.1730	0.1749
CaO	31.76	0.25	0.017832	0.017832	Ca	0.5663	0.5663	5.1949	5.2517
Total						1.7254	2.6164	15.8270	16.0000

With $N_c = 16$ and $N_o = 24$, $S_c = 1.7254$ and $S_o = 2.6164$ we obtain:-

$D = 24 \times 1.7254 / 16 - 2.6164 = -0.0283$ (equation 11.45)

$r_4 = 0.1240 + (2 \times -0.0283) = 0.0674$ (equation 11.47 with q = 4)

$r_5 = 0.0 - (2 \times -0.0283) = 0.0566$ (equation 11.46 with p = 5)

$Fe^{3+} = k_4' = 0.0674 \times 16 / 1.7254 = 0.6250$ (equation 11.48)

$Fe^{2+} = k_5' = 0.0566 \times 16 / 1.7254 = 0.5249$ (equation 11.48)

$Fe_2O_3 = 0.0674 / 0.012524 = 5.38\%$ (equation 11.57)

$FeO = 0.0566 / 0.013919 = 4.07\%$ (equation 11.56)

start with a more generalized expression for the weighted sum of cations W_k, where:-

$$W_k = \sum k_i g_i = N_o \sum x_i a_i g_i / S_o \qquad (11.40)$$

where g_i are weighting factors for each cation. Remembering from equation 11.37 that $S_o = \sum x_i b_i$, we can then obtain:-

$$\frac{\partial W_k}{\partial x_i} = (N_o a_i g_i S_o - b_i N_o \sum x_i a_i g_i) / S_o^2$$

$$= (N_o a_i g_i - b_i W_k) / S_o \qquad (11.41)$$

which can be substituted into equation 11.3 to give:-

$$\text{var } W_k = \sum (N_o a_i g_i - b_i W_k)^2 s_i^2 / S_o^2 \qquad (11.42)$$

where s_i is the standard deviation of the ith oxide (note that throughout section 11.3 the simplified notation for standard deviation is used). This general expression can then be used to calculate the variance of any weighted sum of cations. In particular two special cases should be noted:-

1. When all the weighting factors are zero, except for g_i which is equal to 1, then $W_k = k_i$ and equation 11.42 will give the variance of the ith cation calculated on the basis of N_o oxygens.
2. When all the weighting factors are 1, then $W_k = \sum k_i$ the total number of cations calculated on the basis of N_o oxygens.

It should be noted that the variance of a sum of cations is always less than the sum of the variances of the individual cations. This is due to the fact that x_i occurs in both the numerator and denominator of equation 11.39, so that equation 11.5 cannot be used.

Numerical example

We will now use equation 11.42 to calculate the standard deviation of Si, Fe^{3+} and the total sum of cations on a basis of 24 oxygens using the data of Table 11.7. For clarity the individual values of $N_o a_i g_i$, $b_i W_k$ and $(N_o a_i g_i - b_i W_k)^2 s_i^2$ used in equation 11.42 have been tabulated in Table 11.8. Note that the standard deviation of the total sum of cations is smaller than that of either Si or Fe^{3+} as mentioned above.

11.3.2 Partitioning iron between FeO and Fe_2O_3

Calculations based on mineral stoichiometry are now frequently used to partition iron theoretically into its two valency states. This is especially useful for techniques such as electron microprobe and x-ray fluorescence analysis where only total iron is determined, although the theory presented here is also valid for adjusting analyses in which FeO and Fe_2O_3 have both been determined. Furthermore, the theory can be used for any other elements which may exist in more than one valency state.

TABLE 11.8

Example of the use of equation 11.42 to calculate the standard deviation of a weighted sum of cations on the basis of a fixed number of oxygens, using the grossular garnet data of Table 11.7. The columns show the values of some of the component parts of equation 11.42 which are abbreviated $A = N_0 a_i g_i$ and $B = b_i W_k$. For each example $N_0 = 24$ and $S_0 = 2.6164$ (Table 11.7). P.E. = percentage relative error.

Calculation for Si ($W_k = 5.9066$)

	g_i	A	B	$(A - B)^2 s_i^2$
SiO_2	1.0	0.399435	0.196608	0.002571
TiO_2	0.0	0.000000	0.147852	0.000079
Al_2O_3	0.0	0.000000	0.173789	0.000680
Fe_2O_3	0.0	0.000000	0.110962	0.000493
FeO	0.0	0.000000	0.082211	0.000000
MnO	0.0	0.000000	0.083265	0.000003
MgO	0.0	0.000000	0.146549	0.000174
CaO	0.0	0.000000	0.105325	0.000693
			Sum	0.004692

Standard deviation = $\sqrt{0.004692} / 2.6164 = 0.0262$; P.E. = 0.44

Calculation for Fe^{3+} ($W_k = 1.1373$)

	g_i	A	B	$(A - B)^2 s_i^2$
SiO_2	0.0	0.000000	0.037857	0.000090
TiO_2	0.0	0.000000	0.028469	0.000003
Al_2O_3	0.0	0.000000	0.033463	0.000025
Fe_2O_3	1.0	0.300578	0.021366	0.003118
FeO	0.0	0.000000	0.015830	0.000000
MnO	0.0	0.000000	0.016033	0.000000
MgO	0.0	0.000000	0.028218	0.000006
CaO	0.0	0.000000	0.020281	0.000026
			Sum	0.003268

Standard deviation = $\sqrt{0.003268} / 2.6164 = 0.0218$; P.E. = 1.9

Calculation for total sum of cations ($W_k = 15.8270$)

	g_i	A	B	$(A - B)^2 s_i^2$
SiO_2	1.0	0.399435	0.526821	0.001014
TiO_2	1.0	0.300380	0.396176	0.000033
Al_2O_3	1.0	0.470767	0.465676	0.000001
Fe_2O_3	1.0	0.300578	0.297328	0.000000
FeO	1.0	0.334046	0.220289	0.000000
MnO	1.0	0.338326	0.223112	0.000005
MgO	1.0	0.595468	0.392686	0.000333
CaO	1.0	0.427965	0.282224	0.001328
			Sum	0.002714

Standard deviation = $\sqrt{0.002714} / 2.6164 = 0.0199$; P.E. = 0.13

Consider an analysis of a mineral containing an element that exists in two valency states. Using equation 11.40 we can find the total number of cations on a basis of N_O oxygens as:-

$$\sum k_i = N_O S_c / S_O \tag{11.43}$$

where, S_c is the raw sum of cations as defined by equation 11.38. Obviously, $\sum k_i$ will not be exactly equal to N_c, the total number of cations that should be present for every N_O oxygens according to the mineral formula. However, of we assume that the mineral is stoichiometric, $\sum k_i$ can be made to be exactly equal to N_c by converting cations from one valency state to another. Although this will alter S_O, the raw sum of oxygens, it will not alter S_c, the raw sum of cations. The deficiency of oxygen D, by which the raw sum of oxygen must be increased to make $\sum k_i$ equal to N_c can then be found from the relationship that:-

$$N_c = N_O S_c / (S_O + D) \tag{11.44}$$

which gives:-

$$D = N_O S_c / N_c - S_O \tag{11.45}$$

When the oxygen deficiency, D, is positive, the amount of the more oxidized cation must be increased with respect to the more reduced cation, but when D is negative the more oxidized cation must be reduced in amount. If the pth and qth oxides are the more reduced and oxidized states, respectively, it follows that the theoretical raw amounts of the two cations on a basis of N_O oxygens are given by:-

$$r_p = x_p a_p - FD \tag{11.46}$$

and

$$r_q = x_q a_q + FD \tag{11.47}$$

where F is the number of cations of the more reduced state that must be converted to the more oxidized state to produce 1 oxygen. For the conversion of Fe^{2+} to Fe^{3+} and Ti^{3+} to Ti^{4+}, F = 2. As the proportionality for converting raw cations to cations on a basis of N_O oxygens is now $N_O/(S_O+D)$, which is also equal to N_c/S_c (equation 11.44), it follows that the amount of any cation after the partitioning of an element is given by:-

$$k_i' = N_c x_i a_i / S_c \tag{11.48}$$

and in the special case of the two partitioned cations we have, using equation 11.46, that:-

$$\begin{aligned} k_p' &= N_c r_p / S_c \\ &= N_c(x_p a_p - FD) / S_c \\ &= N_c(x_p a_p - FN_O S_c / N_c + FS_O) / S_c \\ &= N_c(x_p a_p + FS_O) / S_c - FN_O \end{aligned} \tag{11.49}$$

and similarly using equation 11.47 that:-

$$k_q' = N_c(x_q a_q - FS_0) / S_c + FN_0 \tag{11.50}$$

These calculations are also shown in Table 11.7. Again, it is easier to derive the variances of k_p' and k_q' as special cases of the generalized expression for the weighted sum of cations, W_k', calculated after the element has been partitioned. However, the weighted sum is now complicated by the fact that is may or may not contain a partitioned element, so that it is best written as:-

$$W_k' = N_c(x_1 a_1 g_1 + \ldots + r_p g_p + r_q g_q + \ldots + x_n a_n g_n) / S_c \tag{11.51}$$

where n is the number of oxides in the analysis and the g_i are weighting factors as before. Using equations 11.46, 11.47 and 11.45 this can be rewritten as:-

$$W_k' = N_c \left[\sum x_i a_i g_i + F(g_q - g_p)(N_0 S_c / N_c - S_0) \right] / S_c \tag{11.52}$$

Differentiating this with respect to x_i, and remembering the relationships of equations 11.37 and 11.38, we then obtain:-

$$\frac{\partial W_k'}{\partial x_i} = \left[a_i(N_c g_i + F_2) - F_1 b_i \right] / S_c \tag{11.53}$$

where $F_1 = F(g_q - g_p)N_c$ and $F_2 = F_1 N_0/N_c - W_k'$. The variance of a weighted sum of cations after partitioning is then given by:-

$$\text{var } W_k' = \sum \left[a_i(N_c g_i + F_2) - F_1 b_i \right]^2 s_i^2 / S_c^2 \tag{11.54}$$

where s_i is the standard deviation of the ith oxide as before. Several points should be noted about this equation.

1. When all the weighting factors are zero, except g_i which is 1, then $W_k' = k_i'$ and equation 11.54 gives the variance of the ith cation after partitioning. If $i = p$ or q, then the variance of k_i' is always greater than the variance of k_i given by equation 11.42. However, if i is not equal to p or q, then the variance of k_i' may be greater or less than the variance of k_i.
2. If both the partitioned cations appear in the weighted sum with equal weights (i.e. $g_p = g_q$) then $F_1 = 0$ and $F_2 = -W_k'$. Under these conditions equations 11.54 and 11.42 are extremely similar.
3. If all the weighting factors are 1, $W_k' = N_c$, $F_1 = 0$ and $F_2 = -W_k'$, so that var $W_k' = 0$, which, of course, is to be expected as W_k' is the constant N_c.

Numerical example

An example of the partitioning of iron between Fe^{2+} and Fe^{3+} is also given in Table 11.7 for the grossular garnet analysis. Some worked examples of the use of equation 11.54 to calculate the standard deviation of cations after the iron has been partitioned are given in Table 11.9. As expected, the standard deviation of Fe^{3+} (0.0643) is considerably greater than the value obtained before theoretically partitioning the iron (0.0218, Table 11.8).

TABLE 11.9

Example of the use of equation 11.54 to calculate the standard deviation of a weighted sum of cations after the iron has been theoretically partitioned, using the grossular garnet data of Table 11.7. The columns show the values of some of the component parts of equation 11.54 which are abbreviated $A = a_i(N_c g_i + F_2)$ and $B = F_1 b_i$. For each example $N_c = 16$, $N_o = 24$ and $S_c = 1.7254$ (Table 11.7). P.E. = percentage relative error.

Calculation for Si ($W_k' = 5.9712$, $F_1 = 0.0$, $F_2 = -5.9712$)

	g_i	A	B	$(A - B)^2 s_i^2$
SiO_2	1.0	0.166912	0.000000	0.001741
TiO_2	0.0	-0.074734	0.000000	0.000020
Al_2O_3	0.0	-0.117126	0.000000	0.000309
Fe_2O_3	0.0	-0.074783	0.000000	0.000224
FeO	0.0	-0.083110	0.000000	0.000000
MnO	0.0	-0.084175	0.000000	0.000003
MgO	0.0	-0.148151	0.000000	0.000178
CaO	0.0	-0.106477	0.000000	0.000709
			Sum	0.003183

Standard deviation = $\sqrt{0.003183}$ / 1.7254 = 0.0327; P.E. = 0.55

Calculation for Fe^{3+} ($W_k' = 0.6250$, $F_1 = 32$, $F_2 = 47.3750$)

	g_i	A	B	$(A - B)^2 s_i^2$
SiO_2	0.0	0.788469	1.065161	0.004785
TiO_2	0.0	0.592937	0.801013	0.000156
Al_2O_3	0.0	0.929274	0.941534	0.000003
Fe_2O_3	1.0	0.793714	0.601156	0.001483
FeO	0.0	0.659393	0.445395	0.000000
MnO	0.0	0.667842	0.451102	0.000019
MgO	0.0	1.175430	0.793958	0.001179
CaO	0.0	0.844784	0.570620	0.004698
			Sum	0.012323

Standard deviation = $\sqrt{0.012313}$ / 1.7254 = 0.0643; P.E. = 10.3

11.3.3 Weighted sums of oxides

In most cases calculating the variance of a weighted sum of oxides is a trivial exercise and can be solved using equation 11.4 and 11.5. However, if the theoretical oxide values of the two partitioned cations are included in the sum, then a more generalized expression must be considered. In general, a weighted sum of oxides W_x' may be written as:-

$$W_x' = x_i g_i + ... + x_p' g_p + x_q' g_q + ... + x_n g_n \tag{11.55}$$

where the g_i are weighting factors as before, and x_p' and x_q' are the theoretical oxide values of the partitioned cations as given by:-

$$x_p' = r_p / a_p = x_p - FD / a_p \tag{11.56}$$

and

$$x_q' = r_q / a_q = x_q - FD / a_q \tag{11.57}$$

Substituting these values of x_p' and x_q' into equation 11.55, and using equation 11.45, we obtain:-

$$\begin{aligned}
W_x' &= \sum x_i g_i + F_3 D \\
&= \sum x_i g_i + F_3(N_0 S_c / N_c - S_0) \tag{11.58}
\end{aligned}$$

where $F_3 = F(g_q/a_q - g_p/a_p)$. Using equation 11.3, the variance of W_x' is then given by:-

$$\text{var } W_x' = \sum \Big[g_i + F_3(N_0 a_i / N_c - b_i) \Big]^2 s_i^2 \tag{11.59}$$

Again several points should be noticed about this expression.

1. When $g_q = g_p = 0$, $F_3 = 0$, so that $\text{var } W_x' = \sum g_i^2 s_i^2$ which is simply a restatement of equations 11.4 and 11.5. This can be used to find the variance of the oxide total of the original analysis.
2. When all the g_i are equal to 1, including g_p and g_q, then W_x' is the analysis total including the theoretical oxide values of the partitioned cations.
3. When all the g_i are zero except g_p (or g_q) which is 1, $W_x' = x_p'$ (or x_q').

11.3.4 Weighted cation ratios

Some of the most important parameters calculated from mineral analyses are the ratios that form most of the estimates of mineral compositions in terms of end members. In general, both the numerator and denominator of such ratios can be weighted sums of cations which may or may not include partitioned cations and can be generalized as:-

$$\begin{aligned}
V_k' &= \frac{x_1 a_1 g_1 + \cdots + r_p g_p + r_q g_q + \cdots + x_n a_n g_n}{x_1 a_1 h_1 + \cdots + r_p h_p + r_q h_q + \cdots + x_n a_n h_n} \\
&= \frac{\sum x_i a_i g_i + (g_q - g_p)FD}{\sum x_i a_i h_i + (h_q - h_p)FD} \tag{11.60}
\end{aligned}$$

where both the g_i and the h_i are weighting factors. In this case the basis on which the cations were originally calculated is immaterial, as the proportionalities cancel out. Differentiating this expression with respect to x_i, and using equation 11.3, we obtain the variance of the weighted ratio as:-

$$\text{var } V_k' = \sum \Big[a_i(g_i - V_k' h_i) + F_4(N_0 a_i / N_c - b_i) \Big]^2 s_i^2 / \text{Den}^2 \tag{11.61}$$

where $F_4 = \Big[g_q - g_p - V_k'(h_q - h_p) \Big]F$ and Den is the denominator of equation 11.60.

Numerical example

The use of equation 11.61 to calculate the standard deviations of the cation ratios Mg/Fe^{2+} and $Mg/(Mg+Fe^{2+})$, after the iron has been theoretically partitioned, is illustrated in Table 11.10.

11.3.5 Weighted oxide ratios

Although weighted oxide ratios are not common in mineral calculations, they are extremely common as petrological parameters. Again the weighted ratio may or may not include the theoretical oxide calculated from the partitioned cations, so that we can write the general expression:-

$$V_{x'} = \frac{x_1 g_1 + \ldots x_p{}'g_p + x_q{}'g_q + \ldots + x_n g_n}{x_1 h_1 + \ldots x_p{}'h_p + x_q{}'h_q + \ldots + x_n h_n}$$

$$= \frac{\sum x_i g_i + F_3 D}{\sum x_i h_i + F_5 D} \tag{11.62}$$

where F_3 is as defined for equation 11.58 and $F_5 = (h_q/a_q - h_p/a_p)F$. Using equation 11.3, the variance of $V_{x'}$ is then given by:-

TABLE 11.10

Example of the use of equation 11.61 to calculate the standard deviation of a weighted cation ratio, using the grossular garnet data of Table 11.7. The columns show the values of some of the component parts of equation 11.61 which are abbreviated $A = a_i(g_i - V_k{}'h_i)$ and $B = F_4(N_o a_i/N_c - b_i)$. For each example $N_c = 16$ and $N_o = 24$. P.E. = percentage relative error.

Calculation for Mg/Fe^{2+} ($V_k{}' = 0.3332$, $F_4 = 0.6664$)

	g_i	h_i	A	B	$(A + B)^2 s_i{}^2$
SiO_2	0.0	0.0	0.00000000	-0.00554585	0.00000192
TiO_2	0.0	0.0	0.00000000	-0.00417054	0.00000006
Al_2O_3	0.0	0.0	0.00000000	0.00000000	0.00000000
Fe_2O_3	0.0	0.0	0.00000000	0.00000000	0.00000000
FeO	0.0	1.0	-0.00463797	0.00463797	0.00000000
MnO	0.0	0.0	0.00000000	0.00469740	0.00000001
MgO	1.0	0.0	0.02481119	0.00826761	0.00000886
CaO	0.0	0.0	0.00000000	0.00594195	0.00000221
				Sum	0.00001306

Standard deviation = $\sqrt{0.00001306}$ / 0.0566 = 0.0638; P.E. = 19.2

Calculation for $Mg/(Mg+Fe^{2+})$ ($V_k{}' = 0.2499$, $F_4 = 0.4999$)

	g_i	h_i	A	B	$(A + B)^2 s_i{}^2$
SiO_2	0.0	0.0	0.00000000	-0.00415974	0.00000108
TiO_2	0.0	0.0	0.00000000	-0.00312817	0.00000004
Al_2O_3	0.0	0.0	0.00000000	0.00000000	0.00000000
Fe_2O_3	0.0	0.0	0.00000000	0.00000000	0.00000000
FeO	0.0	1.0	-0.00347877	0.00347877	0.00000000
MnO	0.0	0.0	0.00000000	0.00352334	0.00000000
MgO	1.0	1.0	0.01860996	0.00620123	0.00000499
CaO	0.0	0.0	0.00000000	0.00445684	0.00000124
				Sum	0.00000735

Standard deviation = $\sqrt{0.00000735}$ / (0.0189 + 0.0566) = 0.0359; P.E. = 14.4

$$\text{var } V_x' = \sum \left[g_i - V_x'h_i + F_6(N_o a_i / N_c - b_i) \right]^2 s_i^2 / \text{Den}^2 \qquad (11.63)$$

where $F_6 = F_3 - V_x'F_5$ and Den is the denominator of equation 11.62.

Numerical example

The use of equation 11.63 to calculate the standard deviations of the oxide ratios MgO/FeO and MgO/(MgO+FeO) using the theoretical amounts of FeO and Fe_2O_3 given in Table 11.7, is illustrated in Table 11.11.

11.4 NORMATIVE CALCULATIONS

As petrologists make considerable use of norms, it is appropriate to consider the propagation of errors in their calculation. Readers not familiar with the general principles of normative calculations may refer to Johannsen (1950) for descriptions of several types. Although the following discussion is applicable to norms in general the detailed examples are all taken from the CIPW norm, originally described by Cross et al. (1903) and more recently by

TABLE 11.11

Example of the use of equation 11.63 to calculate the standard deviation of a weighted oxide ratio, using the grossular garnet data of Table 11.7. The columns show the values of some of the component parts of equation 11.63 which are abbreviated $A = g_i - V_x'h_i$ and $B = F_6(N_o a_i/N_c - b_i)$. For each example $N_c = 16$ and $N_o = 24$. P.E. = percentage relative error.

Calculation for MgO/FeO ($V_x' = 0.1869$, $F_6 = 26.8606$)

	g_i	h_i	A	B	$(A + B)^2 s_i^2$
SiO_2	0.0	0.0	0.000000	-0.223522	0.003123
TiO_2	0.0	0.0	0.000000	-0.168091	0.000102
Al_2O_3	0.0	0.0	0.000000	0.000000	0.000000
Fe_2O_3	0.0	0.0	0.000000	0.000000	0.000000
FeO	0.0	1.0	-0.186930	0.186930	0.000014
MnO	0.0	0.0	0.000000	0.189326	0.014398
MgO	1.0	0.0	1.000000	0.333221	0.003585
CaO	0.0	0.0	0.000000	0.239487	
				Sum	0.021221

Standard deviation = $\sqrt{0.021221}$ / 4.07 = 0.0358;   P.E. = 19.2

Calculation for MgO/(MgO+FeO) ($V_x' = 0.1575$, $F_6 = 22.6303$)

	g_i	h_i	A	B	$(A + B)^2 s_i^2$
SiO_2	0.0	0.0	0.000000	-0.188319	0.002217
TiO_2	0.0	0.0	0.000000	-0.141618	0.000072
Al_2O_3	0.0	0.0	0.000000	0.000000	0.000000
Fe_2O_3	0.0	0.0	0.000000	0.000000	0.000000
FeO	0.0	1.0	-0.157491	0.157491	0.000010
MnO	0.0	0.0	0.000000	0.159509	0.010220
MgO	1.0	1.0	0.842509	0.280742	0.002544
CaO	0.0	0.0	0.000000	0.201770	
				Sum	0.015063

Standard deviation = $\sqrt{0.015063}$ / (0.76 + 4.07) = 0.0254;   P.E. = 16.1

Cox et al. (1979), Kelsey (1965) and Washington (1917). Unless otherwise mentioned, the abbreviations used for the normative constituents are the same as those used in Table 11.12.

Unfortunately, the problem of the propagation of errors in normative calculations is complicated by the fact that the normative constituents are *discontinuous functions* of the oxides. This means that general expressions cannot be written relating the oxides to the normative constituents without stating strict sets of boundary conditions under which they apply.

The discontinuities are due to two properties of the norm calculation. Firstly, normative constituents cannot be negative, so that they may suddenly disappear on entering certain regions of the p-dimensional oxide space and be replaced by other normative constituents. For example, in a particular region of oxide space there may not be enough SiO_2 to form Q, so that Q will disappear and its place may be taken by Ol as Hy is desilicated. Secondly, the values of many normative constituents are calculated from the smaller of two molecular proportions. For example, if the molecular amount of TiO_2 is less than FeO, the molecular amount of Il is determined by the amount of TiO_2, but if TiO_2 is greater than FeO, the amount of Il is governed by the amount of FeO. So that if norms are calculated for a series of linearly related compositions in which TiO_2 increases at a greater rate than FeO, the amount of Il will change at a constant rate until the amount of TiO_2 is greater than FeO, at which point the rate of change of Il must decrease, thereby causing a discontinuity. The amount of Il is then determined by the amount of FeO. These types of changes can be clearly seen in the CIPW norms of two sets of linearly related compositions in Tables 5.8 and 5.9. The behaviour of Lc is particularly striking in Table 5.9, as it suddenly appears and then disappears.

Before proceeding further we must clarify some terms and properties of normative calculations. A *normative constituent* can be defined as one of the calculated normative mineral species, including a final silica deficiency, or the ratio of the molecular proportions of FeO/(MgO + FeO) used in the calculation of the normative pyroxenes and olivine, and abbreviated throughout as FM. A *normative type* may then be defined as a valid combination of normative constituents as laid down by the rules for calculating the particular norm. Some of the more common CIPW normative constituents and types, found in a sample of 34941 analyses of igneous rocks from the CLAIR and PETROS files, are given in Table 11.12. With these definitions three important properties of normative calculations can then be formalized.

1. *The number of normative constituents in any normative type*
   is equal to p, the number of discrete oxides being used. (11.64)

This is simply an expression of the fact that p transformed variables (the normative constituents) are needed to uniquely define an original set of p variables (the oxides). If less than p transformed variables are used the definition is not unique, so that there are an infinite number of sets of original variables that are capable of producing the same set of transformed

TABLE 11.12

The frequency of occurrence of CIPW normative constituents and types in a sample of 34941 analyses of igneous rocks from the CLAIR and PETROS data files in which the 9 oxides, SiO_2, TiO_2, Al_2O_3, Fe_2O_3, FeO, CaO, MgO, Na_2O and K_2O are present. 122 normative types are found among the 34941 analyses. The abbreviations used for the normative constituents are:- Ab = albite; Ac = acmite; An = anorthite; C = corundum; Cs = dicalcium silicate; Di = diopside; FM = molecular FeO/(MgO + FeO) in olivine and pyroxene; Hm = hematite; Hy = hypersthene; Il = ilmenite; Kp = kaliophilite; Ks = potassium metasilicate; Lc = leucite; Mt = magnetite; Ne = nepheline; Ns = sodium metasilicate; Ol = olivine; Or = orthoclase; Pf = perovskite; Q = quartz; Ru = rutile; SD = final silica deficiency; Tn = titanite (sphene); Wo = wollastonite.

Percentage occurrence of the normative constituents in the 34941 analyses

Q	60.7	C	17.2	Or	97.2	Ab	95.7	An	95.2	Ne	23.2
Lc	4.0	Kp	0.4	Ac	4.8	Ns	1.9	Ks	0.1	Di	82.3
Hy	74.2	Wo	5.7	Ol	36.1	Cs	2.8	Hm	14.6	Mt	95.7
Il	100.0	Tn	1.1	Pf	0.6	Ru	1.0	SD	0.2	FM	85.3

Number of times normative constituents occur in the 122 normative types

Q	27	C	16	Or	86	Ab	54	An	54	Ne	72
Lc	48	Kp	21	Ac	68	Ns	24	Ks	17	Di	83
Hy	30	Wo	33	Ol	72	Cs	32	Hm	67	Mt	60
Il	122	Tn	17	Pf	28	Ru	10	SD	11	FM	46

The 25 most frequent normative types found in the 34941 analyses

Number	%	Normative constituents
12578	36.00	Q - Or Ab An - - - - Di Hy - - - - Mt Il - - FM
5134	14.69	- - Or Ab An - - - - Di Hy - Ol - - Mt Il - - FM
4745	13.58	- - Or Ab An Ne - - - Di - - Ol - - Mt Il - - FM
4188	11.99	Q C Or Ab An - - - - - Hy - - - - Mt Il - - FM
1345	3.85	Q - Or Ab An - - - - Di Hy - - - Hm Mt Il - - -
1129	3.23	Q C Or Ab An - - - - - Hy - - - Hm Mt Il - - -
556	1.59	- - - - An Ne Lc - - Di - - Ol Cs - Mt Il - - FM
413	1.18	- - Or Ab An Ne - - - Di - - Ol - Hm Mt Il - - -
334	0.96	Q - Or Ab An - - - - Di - Wo - - Hm Mt Il - - -
323	0.92	- - Or Ab An Ne - - - Di - Wo - - - Mt Il - - FM
287	0.82	Q - Or Ab An - - - - Di - Wo - - - Mt Il - - FM
287	0.82	Q C Or Ab An - - - - - Hy - - - Hm - Il - Ru -
279	0.80	- - Or - An Ne Lc - - Di - - Ol - - Mt Il - - FM
271	0.78	- - Or Ab An Ne - - - Di - Wo - - Hm Mt Il - - -
267	0.76	Q - Or Ab - - - Ac Ns Di Hy - - - - - Il - - FM
248	0.71	Q - Or Ab - - - Ac - Di Hy - - - - Mt Il - - FM
215	0.62	Q - Or Ab An - - - - Di Hy - - - Hm - Il Tn - -
205	0.59	- - Or Ab An - - - - Di Hy - Ol - Hm Mt Il - - -
178	0.51	- - Or Ab - Ne - Ac Ns Di - - Ol - - - Il - - FM
171	0.49	- C Or Ab An - - - - - Hy - Ol - - Mt Il - - FM
129	0.37	- - Or Ab - Ne - Ac - Di - - Ol - - Mt Il - - FM
124	0.35	- - - - An Ne Lc - - Di - - Ol Cs Hm Mt Il - - -
112	0.32	- - Or Ab - Ne - Ac - Di - Wo - - - Mt Il - - FM
102	0.29	- C Or Ab An Ne - - - - - - Ol - - Mt Il - - FM
100	0.29	- - Or Ab - Ne - Ac Ns Di - Wo - - - - Il - - FM

Of the remaining 97 normative types, 5 are represented by 53 to 78 analyses each 30 are represented by 11 to 47 analyses each, 10 are represented by 5 to 9 analyses each and 52 are represented by only 1 to 4 analyses each.

variables (cf. Le Maitre, 1965). Similarly, if more than p are used, certain of the transformed variables must contain redundant information. This property is a useful check on the logic of any norm program and can be verified in Tables 5.8, 5.9 and 11.12. However, care must be taken not to count oxides such as MnO, SrO, Rb_2O, NiO, CoO etc. as discrete oxides, as these are normally added to other oxides. Similarly, the normative constituents for olivine and pyroxenes should be chosen to minimize their number so that, for example, Ol, Hy, Di and FM should be used rather than Fo (forsterite) and Fa (fayalite) for the olivine, Wo, En (enstatite) and Fs (ferrosilite) for the clinopyroxene and En and Fs for the orthopyroxene. The redundancy in using 7 normative constituents is obvious when it is realized that the Di is always calculated stoichiometrically, so that given En and Fs the amount of Wo is fixed. Similarly, as FM has the same value for olivine and the pyroxenes, only Fo (or Fa) and En (or Fs) together with FM, are necessary to define their compositions.

2. *Within a single normative type, all the normative constituents,*
   except FM, vary linearly with the oxides.   (11.65)

This means that for each normative type a set of linear equations, relating the normative constituents to the oxides, can be written by performing the norm calculation algebraically. However, in general, these linear equations will be different for each normative type. For example, the Differentiation Index, Q+Or+Ab+Ne+Lc+Kp (Thornton & Tuttle, 1960), is a function of only Na_2O and K_2O in normative types containing Hy and Ol, but for Q- and Ne-bearing normative types, it is a function of all the oxides (Le Maitre, 1968, p.228). This is one of the disadvantages of projecting oxide data into normative space as one is using, in effect, a series of discontinuous funtions for the projection (Le Maitre, 1976b).

3. *Within a single normative type, FM either varies*
   curvilinearly with the oxides or is zero.   (11.66)

This is because FM is always a ratio of two linear functions of the oxides. However, if all the FeO has been allotted to Il and Mt, so that none is left for the olivine and pyroxenes, then FM is always zero and does not count as a normative constituent.

Now to return to the propagation of errors. As it is not possible to write expressions for normative constituents that are valid for all normative types, the errors must be assigned at each step in the calculation – a procedure that is ideally performed by a computer program. As the errors are cumulative, it also follows that those normative constituents which are allocated first will have smaller errors than those allocated later from residual amounts of the oxides. Fortunately, the theory is simple and can be implemented in the following manner.

1. Obtain var x_i, the variances of the weight percent oxides, x_i, by either using equation 11.22 or 11.23, or by assuming values. Note that in this section the

notation for variance is different from that used in section 11.3.

2. If only total iron has been determined and it is desired to adjust the values of x_{FeO} and $x_{Fe_2O_3}$, then their variances must also be adjusted. If the ratio of $x_{FeO}/(x_{FeO} + x_{Fe_2O_3})$ is to be made equal to a constant, say R, then using equations 11.4 and 11.5 we have:-

$$\text{var } x'_{FeO} = R^2(\text{var } x_{FeO} + \text{var } x_{Fe_2O_3}) \tag{11.67}$$

and

$$\text{var } x'_{Fe_2O_3} = (1 - R)^2(\text{var } x_{FeO} + \text{var } x_{Fe_2O_3}) \tag{11.68}$$

where the primes indicate the new values. However, if one of the values is set to a constant, say $x_{Fe_2O_3}$ is set to 1.5%, then it follows that the new variances would be:-

$$\text{var } x'_{FeO} = \text{var } x_{FeO}$$
$$\text{var } x'_{Fe_2O_3} = 0 \tag{11.69}$$

3. The variances of the initial molecular proportions, w_i, are then calculated using equation 11.4 as:-

$$\text{var } w_i = \text{var } x_i / (\text{Molecular weight ith oxide})^2 \tag{11.70}$$

4. If z_k, the molecular proportion of the kth normative constituent, is initially calculated as the smaller of two molecular proportions, w_i and w_j, as for example with Mt, Il, Or, Ab and An, then two conditions must be considered. When $w_i > w_j$ it follows that:-

$$\text{var } z_k = \text{var } w_j \tag{11.71}$$

The remaining amount of the ith oxide, w_i', is then set to $w_i - w_j$, so that its variance is given by equation 11.5 as:-

$$\text{var } w_i' = \text{var } w_i + \text{var } w_j \tag{11.72}$$

However, when $w_i < w_j$ it follows that:-

$$\text{var } z_k = \text{var } w_i \tag{11.73}$$

and the remaining amount of the jth oxide, w_j', is set to $w_j - w_i$, so that its variance is given by:-

$$\text{var } w_j' = \text{var } w_j + \text{var } w_i \tag{11.74}$$

Note that the variance of the remaining amount of the larger oxide is independent of which oxide is the larger, i.e. equations 11.72 and 11.74 give the same values.

5. If a molecular proportion w_i' is in excess and it is to be allotted entirely to the kth normative constituent as, for example, with C, Ru, Hm and Wo, it then follows that:-

$$\text{var } z_k = \text{var } w_i' \tag{11.75}$$

6. The variance of FM, the molecular ratio of $w_{FeO}/(w_{MgO} + w_{FeO})$ used in the calculation of the pyroxenes and olivine, can be evaluated using equation 11.3 as:–

$$\text{var FM} = (w_{MgO}^2 \text{ var } w_{FeO} + w_{FeO}^2 \text{ var } w_{MgO}) / (w_{MgO} + w_{FeO})^4 \quad (11.76)$$

Remember that w_{FeO} is the molecular proportion of FeO remaining at the time of formation of the provisional pyroxenes.

7. Now we come to the allocation of silica. The molecular amount of excess silica, z_Q, to form normative Q, is given by the original amount of silica, w_{SiO_2}, less the amount of silica required for the provisionally allotted normative constituents. This can be expressed as:–

$$z_Q = w_{SiO_2} - \sum g_i z_i \quad (11.77)$$

where g_i is the number of molecules of SiO_2 in the ith normative constituent formula, e.g. for Or and Ab it is 6, for An and Di it is 2, etc. It follows, therefore, that the variance of z_Q is given by:–

$$\text{var } z_Q = \text{var } w_{SiO_2} + \sum g_i^2 \text{ var } z_i \quad (11.78)$$

In many cases, z_Q in equation 11.77 is negative, so that following the CIPW rules we must start to desilicate some of the provisionally allotted normative constituents. If, however, we proposed a norm in which undersaturation was reported as a negative amount of normative Q, then equations 11.77 and 11.78 would still be valid.

8. As the formulae for calculating the amounts of the final sets of normative constituents during desilication are very specific, only the two common assemblages of Hy-Ol and Ab-Ne (Table 11.12) will be treated in detail. Any other types can easily be worked out by analogy. For the Hy-Ol assemblage the formulae for calculating the molecular amounts of Hy and Ol can be written as:–

$$z_{Hy} = 2w_{ASiO_2} - w_{F+M} \quad (11.79)$$
and
$$z_{Ol} = 2(w_{F+M} - w_{ASiO_2}) \quad (11.80)$$

where w_{F+M} and w_{ASiO_2} are the molecular proportions of FeO+MgO and SiO_2, respectively, available to form Hy and Ol. The variances are, therefore, given by:–

$$\text{var } z_{Hy} = 4\text{var } w_{ASiO_2} + \text{var } w_{F+M} \quad (11.81)$$
and
$$\text{var } z_{Ol} = 4(\text{var } w_{ASiO_2} + \text{var } w_{F+M}) \quad (11.82)$$

The variance of the available amount of silica is given by the equation:–

$$\text{var } w_{ASiO_2} = \text{var } w_{SiO_2} + \sum g_i^2 \text{var } z_i \quad (11.83)$$

summed over all the provisionally allocated silicates, except the one being desilicated. This is basically the same as equation 11.78. For the desilication of Ab, the formulae for the allocation of Ab and Ne may be written as:-

$$z_{Ab} = 0.25w_{ASiO_2} - 0.5w_{Na_2O} \tag{11.84}$$

and

$$z_{Ne} = 1.5w_{Na_2O} - 0.25w_{ASiO_2} \tag{11.85}$$

where w_{Na_2O} is the molecular amount of Na_2O available to form Ab and Ne, and w_{ASiO_2} is as before. The variances of Ab and Ne are, therefore, given by:-

$$var\ z_{Ab} = 0.0625var\ w_{ASiO_2} + 0.25var\ w_{Na_2O} \tag{11.86}$$

and

$$var\ z_{Ne} = 2.25var\ w_{Na_2O} + 0.0625var\ w_{ASiO_2} \tag{11.87}$$

where $var\ w_{ASiO_2}$ is calculated using equation 11.83. Similar expressions may be written for the variances of all other sets of desilicated normative constituents.

9. The final step of the norm calculation is to convert the molecular proportions of the normative constituents to weight percentages. As this is done by multiplying the molecular proportions by the molecular weights of the respective normative constituents, it follows that their variances are given by:-

$$var\ Z_i = var\ z_i\ (\text{molecular weight of ith normative constituent})^2 \tag{11.88}$$

where Z_i is the weight percent of the ith normative constituent. Remember that for MgO- and FeO-bearing normative constituents the molecular weight, MWZ_i, is given by:-

$$MWZ_i = MWMgZ_i(1 - FM) + MWFeZ_i(FM) \tag{11.89}$$

where $MWMgZ_i$ and $MWFeZ_i$ are the molecular weights of the Mg and Fe end-members, respectively. Using Table A12 in the Appendix, we can then write for Di that:-

$$\begin{aligned} MWZ_{Di} &= 216.5534\ (1 - FM) + 248.0954FM \\ &= 216.5534 + 31.5420FM \end{aligned} \tag{11.90}$$

Similarly, for Hy and Ol we obtain:-

$$MWZ_{Hy} = 100.3892 + 31.5420FM \tag{11.91}$$

and

$$MWZ_{Ol} = 70.3468 + 31.5420FM \tag{11.92}$$

Note that this expression for the molecular weight of Ol must be used when the molecular amount of Ol is measured in units of FeO+MgO and not SiO_2. Strictly speaking, these molecular weights should not be used in equation 11.88 as they are not constants, as FM is subject to error (equation 11.76). However, the effect is so small that it can be neglected.

Numerical example

Table 11.13 gives a worked example of the calculations of the variances of the normative constituents using a simplified version of the x-ray fluorescence basalt analysis given in Table 11.4, from which the standard deviations are also taken. As only total iron is given, it will be assumed that Fe_2O_3 is 1%, so that its standard deviation is zero (equation 11.69). The initial molecular proportions, w_i, are calculated by dividing x_i, the weight percents of the oxides, by their respective molecular weights obtained from Table A9. For clarity the results have been multiplied by 10^5 and presented as integers, so that, for example, the molecular proportion of SiO_2 is:-

$$w_{SiO_2} = 52 \times 10^5 / 60.0848 = 86544$$

Similarly, var w_i, the variances of the initial molecular proportions, obtained from equation 11.70 are multiplied by 10^{10}, so that, for example, the variance of the molecular proportion of FeO is:-

$$var\ w_{FeO} = 0.01^2 \times 10^{10} / 71.8464^2 = 194$$

The molecular amount of Mt, z_{Mt}, is then calculated as the smaller value of $w_{Fe_2O_3}$ and w_{FeO}, so that we have $z_{Mt} = 626$ and var $z_{Mt} = 0$, (equation 11.71). The remaining amount of FeO, w_{FeO}', is then 11135 - 626 = 10509 and its variance, var w_{FeO}' is given by equation 11.72 as 194 + 0 = 194. The allocation of Ab, An and Di then follow similar procedures.

Before forming provisional Hy the variance of FM may be calculated using equation 11.76 as:-

$$var\ FM = \left[(19849^2 \times 194) + (10509^2 \times 5540)\right] / (10509 + 19849)^4$$
$$= 0.000000810$$

so that the standard deviation of FM is $\sqrt{0.000000810} = 0.0009$. Note that in this example the multiplication factors of 10^5 and 10^{10} cancel out.

After the provisional allocation of Hy, the molecular amount of normative Q is calculated, using equation 11.78, as:-

$$z_Q = 86544 - 29040 - 19742 - 15922 - 22397 = -557$$

As this is negative, we set Q to zero and start to desilicate some of the provisionally allocated normative constituents. In this particular example we have to go no further than the partial desilication of Hy, so that we can use equations 11.79 and 11.80 to determine the molecular amounts of Hy and Ol as:-

$$z_{Hy} = 2 \times 21840 - 22397 = 21283 \quad and$$
$$z_{Ol} = 2\ (22397 - 21840) = 1114$$

The variance of the available amount of SiO_2 is then calculated, using equation 11.83, as:-

$$var\ w_{ASiO_2} = 13573 + (36 \times 2343) + 4\ (3882 + 5154) = 134065$$

TABLE 11.13

Numerical example of the propagation of errors in a CIPW norm calculation, using a simplified version of the x-ray fluorescence basalt analysis given in Table 11.4, from which the standard deviations (s_i) are also taken. For clarity the molecular proportions, w_i, and their variances, var w_i, have been multiplied by 10^5 and 10^{10}, respectively, and presented as integers. The molecular weights used to calculate the w_i are from Table A9. The abbreviations used for the normative constituents are the same as used in Table 11.12.

	SiO_2	Al_2O_3	Fe_2O_3	FeO	MgO	CaO	Na_2O		Wt%	s_i
x_i	52.00	15.00	1.00	8.00	8.00	10.00	3.00			
s_i	0.07	0.04	0.00	0.01	0.03	0.02	0.03			
w_i	86544	14711	626	11135	19849	17832	4840			
var w_i	13573	1539	0	194	5540	1272	2343			
Allocation of magnetite										
z_{Mt}	–	–	626	626	–	–	–	Mt	1.45	0.0
var z_{Mt}	–	–	0	0	–	–	–			
w_i'	–	–	0	10509	–	–	–			
var w_i'	–	–	0	194	–	–	–			
Allocation of provisional albite										
z_{Ab}	29040	4840	–	–	–	–	4840	Ab	25.38	0.25
var z_{Ab}	–	2343	–	–	–	–	2343			
w_i'	57504	9871	–	–	–	–	0			
var w_i'	–	3882	–	–	–	–	0			
Allocation of anorthite										
z_{An}	19742	9871	–	–	–	9871	–	An	27.46	0.17
var z_{An}	–	3882	–	–	–	3882	–			
w_i'	37762	0	–	–	–	7961	–			
var w_i'	–	0	–	–	–	5154	–			
Calculation of molecular ratio of MgO/(MgO+FeO)										
FM	–	–	–	10509	19849	–	–	FM	.3462	.0009
At this stage available FeO is added to MgO and treated as one oxide										
$^wFeO+MgO'$	–	–	–	–	30358	–	–.			
Allocation of provisional diopside										
z_{Di}	15922	–	–	–	7961	7961	–	Di	18.11	0.16
var z_{Di}	–	–	–	–	5154	5154	–			
w_i'	21840	–	–	–	22397	0	–			
var w_i'	–	–	–	–	10888	0	–			
Allocation of provisional hypersthene and quartz										
z_{Hy}	22397	–	–	–	22397	–	–			
z_Q	-557	–	–	–	–	–	–			
Desilication of hypersthene into hypersthene and olivine										
z_{Hy}	21283	–	–	–	21283	–	–	Hy	23.69	0.82
var z_{Hy}	–	–	–	–	547148	–	–			
z_{Ol}	557	–	–	–	1114	–	–	Ol	0.91	0.62
var z_{Ol}	–	–	–	–	579812	–	–			

so that we may then use equations 11.81 and 11.82 to obtain the variances of the molecular amouts of Hy and Ol as:-

$$\text{var } z_{Hy} = 4 \times 134065 + 10888 = 547148 \quad \text{and}$$
$$\text{var } z_{Ol} = 4 (134065 + 10888) = 579812$$

The final stage is to convert the molecular amounts of the normative constituents to weight percents and to use equation 11.88 to obtain their variances. For example, the variance of Ab is given by:-

$$\text{var Ab} = 2343 \times 524.4490^2 / 10^{10} = 0.06444$$

which gives a standard deviation of 0.25. For Hy we must first calculate the appropriate molecular weight using equation 11.91 which gives:-

$$\text{MWZ}_{Hy} = 100.3892 + 31.5420 \times 0.3462 = 111.3090$$

The variance of the weight percent of Hy is then given by equation 11.88 as:-

$$\text{var Hy} = 547148 \times 111.3090^2 / 10^{10} = 0.68$$

so that its standard deviation is 0.82. A similar procedure is then followed for Ol.

As a final check on the calculation of the norm, note that the number of normative constituents is 7, which is the number of discrete oxides used in the example (relationship 11.64), and that the sum of the normative minerals is 97.0, which is the same as the sum of the oxides. The latter is perhaps the best test of any norm calculation and, if performed by computer, the two totals should not differ by more than a unit in the 3rd or 4th decimal place, depending upon the word length of the computer being used.

11.4.1 *Limitations on use*

Finally a word of caution concerning the use of the standard deviations of the normative constituents. If we assume that the oxide determinations are normally distributed, then the values of the normative constituents will, in general, also be normally distributed, so that the methods of Chapter 2 can then be used to make probability statements about the population values of the normative constituents.

However, if the analysis is near to the boundary of a normative type, some of the population distributions of the normative constituents may be effected by the discontinuities across the boundary. They may be either truncated at the boundary, or consist of different normal distributions either side of the boundary. For example, in Table 11.13, the 95% confidence limits of Ol are $0.91 \pm 1.96 \times 0.62$ or 2.13 and −0.31, which means that the population distribution of Ol must be truncated at zero, as the norm calculation does not allow negative normative constituents. Now although we can still use these confidence limits to conclude that there is evidence to suggest that the norm may contain Q, it is difficult to comment on how much Q may be present, as −1% of Ol does not convert to 1% of Q.

Now as soon as Q appears in the norm, the variance of Hy is controlled by var w_{F+M}, the variance of the available amount of FeO + MgO, and not by $4var\ w_{ASiO_2} + var\ w_{F+M}$ (equation 11.81). The population distribution of Hy, therefore, must consist of truncated portions of two normal distributions, with very different standard deviations, joined across the boundary between the two normative types. For example, in Table 11.13, the standard deviation of Hy is 0.82 on the Ol side of the boundary, but is $\sqrt{10888}$ x 111.3090 / 10^5 = 0.12 on the Q side of the boundary. Such a distribution is difficult to use, as the location of the boundary within the distribution is uncertain.

In the particular example used in Table 11.13, the other normative constituents are well away from normative type boundaries, so that no problems arise through the use of their standard deviations.

CHAPTER 12

PETROLOGICAL DATA MANAGEMENT

12.1 INTRODUCTION

One of the problems facing petrologists today, is how to manage the tremendous amount of published and unpublished information that is available. This is particularly true of analytical data. As a result petrologists are turning more and more to computers to store, retrieve and process their data. Part of the impetus for this comes from the increased use of computers in many of todays analytical techniques, as it is often an easy step to produce the analytical results in machine readable form for input to another computer.

A serious consequence of this data explosion is that scientific journals are no longer able to publish all the raw data that is produced and often only publish averages, or plots of the data. As a result, a valuable resource is in danger of being lost. Admittedly, many articles state that photocopies of the raw data may be obtained from the author, or from one of the data storage organizations, but this is not entirely satisfactory, as authors move, and data storage organizations have been known never to have received the data. Such a method of storing data also has the major drawback that it assumes that *one knows that the data exist* before they can be obtained.

In the near future, it is hoped that such information will be available from one of several international databases, to which authors would send their raw data, perhaps even as a condition of publication. Researchers requiring data could then contact the appropriate organization (perhaps a World Data Centre) to request information of a certain type, *without prior knowledge that the data exist*. This may not be as far-fetched as it sounds, as several databases like this already exist. Such databases, of course, have enormous potential, especially in the field of exploration where summaries of the characteristics of rock types and regions are often required at short notice.

12.2 PETROLOGICAL DATABASES

In this book the term *database* will be used in its broad sense to mean any set of data which may be processed by a computer. The way in which the data is physically organized within the database is immaterial. Databases may also consist of one or more *data files*, which are sets of data organised as a physical entity. A database, therefore, may be likened to a library consisting of one or more books, where the books are analagous to data files. As most petrological databases consist of only one data file, the two terms are often used synonymously in the petrological literature.

Before looking at some of the existing types it is important to clarify one or two points. Firstly, databases are not the universal elixir of knowledge. They are only as useful as the information they contain, and as powerful as the computer programs that process them. Secondly, in general they cannot be used to do anything that a human-being with a pencil and paper could not do, *given sufficient time*. Basically, all they provide is an extremely rapid retrieval of data. The use to which the data is then put, and its interpretation, is still in the hands of a human-being, so that the old computer saying "garbage in - garbage out" is still very pertinent.

Databases can be divided broadly into two groups. One type, called *reference databases* (also *bibliographic databases*), indicates where the required information may be found, while the other type, known as *source databases* (also *non-bibliographic* and *numeric databases*), provides the required data. Note that although many source databases do contain numeric data, it is not an essential part of their definition.

Burk (1981), in a recent survey of Geoscience databases which have been described in the literature, indicates that there are at least 46 public reference databases and 82 source databases, of which 11 of the source databases are in the fields of petrology and geochemistry. The actual number of databases in existence must, of course, be considerably greater.

12.2.1 *Some source databases*

We will now look briefly at three source databases that are of general interest to igneous petrologists. They all consist of a single data file and were originally compiled by individual petrologists as specific research projects. In order of creation they are:-

1. RKNFSYS (Chayes, 1972), which contains approximately 16,000 analyses of Cenozoic volcanic rocks from all parts of the world. Stored with each analysis is the geographic location and Tröger number of the rock type (Tröger, 1969). Although it has never been distributed for use elsewhere, it has had many uses as data retrievals and reductions have always been available on request (see the Annual Reports of the Director of the Geophysical Laboratory since 1968-69).

2. CLAIR (Le Maitre, 1973), which consists of approximately 26,000 analyses of plutonic and volcanic rocks of all ages from all parts of the world. Stored with each analysis are trace elements, if determined; latitude and longitude; coded geographic location; rock name as given by the author and the reference from which the data came. Although the database has never been made freely available, copies have been installed elsewhere by special arrangement. It has been extensively used to investigate problems associated with the classification of igneous rocks (Le Maitre, 1976a, 1976b, 1976c, Streckeisen & Le Maitre, 1979), which has been of help to the I.U.G.S. Subcommission on the Systematics of Igneous rocks, as well as

providing numerical examples for this and other books.

3. PETROS (Mutschler et al., 1976, 1978), which now contains over 35,000 analyses of plutonic and volcanic rock analyses from all over the world and is freely available for the cost of reproduction (Mutschler et al., 1978). It contains the same type of information as CLAIR, except that the rock names are stored as raw text and not coded, which makes retrieval by rock name more difficult (see section 12.2.3). This data base is now being used by over 150 organizations (F. Mutschler, per. comm.) and results of its application to exploration problems are beginning to appear in the literature (Mutschler et al., 1981).

Before leaving these 3 databases it is interesting to compare how they differ in their coverage of information. A comparison of RKNFSYS and the volcanic rocks from CLAIR in 1971, revealed that only 15% of the analyses were common to both databases (Chayes & Le Maitre, 1972a, 1972b), a fact which surprised both authors as they had both believed that a considerable amount of the available data had been collected. Assuming that both databases were random samples from the literature, it was estimated that at least 100,000 analyses had already been published in the literature. Similarly, a comparison between PETROS (when it contained 26,000 analyses) and CLAIR in 1976, showed that only 18% of the analyses were in common, which resulted in a merged version of CLAIR and PETROS containing just over 41,000 analyses.

Published maps of the geographical distribution of data from CLAIR (Le Maitre, 1973) and PETROS (Mutschler et al., 1978) also show some differences. Compared with CLAIR, PETROS contains very little data from Europe and Australasia, but far more from the western U.S.A. and the ocean floors. The geographic distribution of data from RKNFSYS has not been published in map form.

IGBA, which was started in 1977 as I.G.C.P. Project No. 163 (Chayes, 1979), is potentially the most useful of all the petrological databases. It is currently under construction and is an attempt to improve on the existing databases by including far more information with each analysis e.g. textures, field occurrence, modal and mineralogical data, etc. The project is international in scope with working groups from 12 countries actively contributing data. As a result, it is hoped to avoid further duplication of effort that occurred during the construction of RKNFSYS, CLAIR and PETROS. Eventually, this database will supercede the others in both number of analyses and content, and will be available to all interested petrologists.

Possibly the largest source database in current use is RASS (van Trump & Miesch, 1977), which is run by the U.S. Geological Survey to maintain their geochemical and petrological data. However, it is not freely available and its contents are restricted mainly to the U.S. Burk (1981) also notes that national source databases exist in Canada (Geochemical Data System), France (GEOGEN & GUF) and Great Britain (NGDB) and mentions some topic orientated source databases of possible interest to petrologists, i.e. two on the geochemistry of manganese nodules, one on lead isotopes and one on volcanology. Notably absent in the literature so far, is any mention of large machine-readable mineralogical, metamorphic or sedimentary databases that would be of interest to petrologists, in spite of the large amount of published information already available in these fields (see, however, Horder, 1981).

12.2.2 Some reference databases

Although there are no specific petrological reference databases, the field of petrology and geochemistry is adequately covered. Five of the databases are international is scope and cover all geological fields. In alphabetical order they are:- GEOARCHIVE (U.K.), GEOREF (U.S.A.), JICIST (Japan), PASCAL-GEODE (France) and VINITI (U.S.S.R). Of these, GEOARCHIVE and GEOREF are available through libraries in many parts of the world via satellite communications. Four other countries also maintain national reference databases, i.e. Australia (AESIS), Canada (GEOSCAN), Czechoslovakia (GEO-INDEX) and West Germany (Geological Science Literature). It must be emphazied that with all of these databases it is very important to study the relevent thesaurus of keywords before requesting information, in order to avoid the extraction of unwanted references.

12.2.3 Starting a database

A large number of articles have been written about Geoscience databases in the last decade (e.g. Clark, 1976; Gill et al., 1977; Gilliland & Grove, 1973; Goubin, 1978; Grandclaude, 1976a, 1976b; Horder, 1981; Hubaux, 1969; Le Maitre, 1973, 1978; Loudon, 1969; Robinson, 1970), but much of the comment has been of a theoretical nature. Here then are some practical comments on how to begin to organize data for use as a simple database.

Firstly, spend a considerable amount of thought on deciding what types of information are to be included in the database. As one usually has a limited amount of time and money to spend on such a project, there is always a compromise between obtaining a large amount of information about a few things, or a small amount of information about many things. You must, of course, include everything that is required for the current project, but some of the items that "may be useful" or "would be nice to include" are probably best left out, unless you have plenty of time, money or help.

With alphanumeric data, e.g. rock names, references, etc., a decision must be made on how the information is to be stored. One method is to store the data as *raw text* as it is written so that, for example, the term "porphyritic olivine basalt" might appear as a string of 26 characters in the alphanumeric data assuming, of course, that enough space has been allotted for storing a rock name of this size. If enough space has not been allotted, then abbreviations have to be used which can lead to problems if they are not used consistently. For example, to retrieve basalts from such a database one might search for "basalt" in the appropriate range of characters allotted to the rock name. This would certainly retrieve many of the basalts, but would exclude abbreviations such as "bas.", and would include terms such as "trachybasalt" and "basaltic andesite" which may not be required. To achieve the desired retrieval the search would have to look for "basalt", not preceded by "trachy" and not followed by "ic", and for all the known abbreviations, such as "bas.". So that, unless one knows the complete vocabulary that has been used, the logic of retrieving from raw text data can be extremely complex. For example, the rock names in PETROS are stored in this way, so that when PETROS was merged with CLAIR, the names had to be changed into the coded form used in CLAIR. This required PETROS to be read about 10 times, with the easier names being decoded first and leaving the more uncommon names, abbreviations and spelling mistakes until last, when they could be looked at individually. Finally, one can never be quite sure that an abbreviation such as "ol.bas." does not stand for olivine basanite, unless one looks at the original source of the data. If, therefore, raw text is to be stored and space is limited, it is strongly recommended that a thesaurus or vocabulary of acceptable terms is used, in order to avoid these problems.

If one is prepared to accept a more compilcated data structure, where each piece of information is separated from the others by special characters, known as *delimiters*, then abbreviations need not be used as each piece of information can be of variable length. However, although locating the start of the information in such a file is somewhat slower, the logic of retrieval without abbreviations is somewhat faster. This method can also conserve space in a data file if many of the items have missing information, as blanks do not have to be stored. A major problem with this type of structure is that if some of the delimiters are accidentally missing or if extra ones are present, it is extremely difficult to decode the record.

In many ways the simplest solution is to use a *code* which is stored in a fixed location within the file, as in CLAIR and certain parts of PETROS. Although this method may require more storage, it greatly simplifies the task of data retrieval, especially if a hierarchical type of code is used. For example, in CLAIR all the rock names are allotted 4 letters, the first being either V, for a volcanic rock, or P, for a plutonic rock. The second letter indicates the pigeon-hole in the mineralogical QAPF classification (Streckeisen, 1978) into which the rock should fall, while the last two letters are simply mnemonics for the rock name, so that we finish up, for example, with VKBA for BAsalt and VKAN for ANdesite. With such a system it is extremely easy to extract, for example, all the volcanic or plutonic rock names. However, the coding does makes the initial work of assembling the data more time consuming and increases the possibility of making errors, unless they are easily remembered mnemonics. These disadvantages, however, are generally outweighed by the simpler logic and increased speed of retrieval. A similar type of hierarchical code has been proposed by Harrison & Sabine (1970), for mineralogical and petrological names.

When codes have been used it is important to check their validity against a master check list. This is only a partial check, however, as it will not exclude the possiblilty that the wrong code was chosen, or that a misspelt code is still a valid code. To maximize the detection of this type of error, it is wise to program the computer to print out any coded data in a readable decoded form by using look-up tables or dictionaries of valid terms.

If major element analyses are being used, it is wise to record the published totals so that the computer can check these against the actual totals. Any discrepancies must then be due to either clerical errors or errors in the published totals. This is virtually the only reliable check that can be made automatically on the validity of petrological data, as the chances of making a mistake in data entry and still obtaining the correct total are extremely small. As a guide approximately 13% of all the analyses in the CLAIR database had incorrect published totals (Le Maitre, 1973).

With data collected from the literature, the possibility of obtaining duplicate sets of data will always exist if secondary sources are used. To eliminate this possibility, it is best to keep the data stored in a sorted order, e.g. in descending order of SiO_2, TiO_2 and Al_2O_3, so that duplicates will occur adjacent to each other within the file. Again as a guide, only two analyses were found in the CLAIR database with identical values of SiO_2, TiO_2 and Al_2O_3 but genuinely different values of all the other oxides.

As the final step, always produce a hard copy of all the data in an easily readable format, with any coded information decoded, for visual checking before accepting the data as correct. Remember that the most important thing about any database is that it must contain accurate data; a large database full of inaccurate data would be worthless.

Once the raw data have been converted into machine readable form, such as punched cards or card images on disc, a potential database exists - all that remains to be done is to acquire the computer programs to process the data.

12.3 DATA PROCESSING PACKAGES

With simple databases, *ad hoc* programs are usually adequate to process the data, but with large sets of complex data it is more efficient to process the data with a series of compatible programs, often called *data processing packages* or *systems*. There are a several geologically orientated data processing packages available, e.g. the CLAIR data processing system (Le Maitre & Ferguson, 1978); GEOIC (Parker, 1981); G-EXEC (Jeffery & Gill, 1976, 1977); HARDROCK (Till, 1977); KEYBAM (Barr et al., 1977); PETPAK (Fitzgerald & Mackinnon, 1977); SIGMA (Kremer et al., 1976); STATPAC (van Trump & Miesch, 1977), all of which have their advantages and disadvantages. While it is not intended to embark upon a critical analysis of each individual package, it is important to discuss some of the attributes that a good data processing package should have.

Two of the most important characteristics of a good data processing package are *flexibility* and *ease of use*. This is particularly true in a research environment where every problem undertaken seems to be different from the last. Again there is often a compromise between these two characteristics, as generally the easier a package is to use, the less flexible it is. Before most packages can be used, some type of language and syntax usually has to be learnt. In some, the language is peculiar to the particular package, e.g. G-EXEC, SIGMA, STATPAC and SPSS (Statistical Package for Social Sciences, Nie et al., 1975) and can be quite complex. It would seem logical, therefore, that if one has to go to the trouble to learn a language and its syntax, one might as well learn one that is going to be useful for other applications, such as Fortran, and use a package like CLAIR. One of the great advantages of such a package is that, by giving the user the option of supplying Fortran subroutines to manipulate the data, the flexibility is limited only by what can be written in Fortran and not by what the designer of the package dictates. Furthermore, any existing Fortran subroutines, including mathematical and statistical packages like IMSL (International Mathematical and Statistical Libraries Inc.), can be used in conjunction with CLAIR.

Some data processing packages give the user the option of deciding how the data is to be physically organized within the database. Basically, there are two extremes ranging from a simple *sequential file*, in which every set of data must be read to get to the last, to a *random access file*, in which indexes are kept indicating where particular sets of data may be found. This provides very quick retrieval when there is a index for the information requested, and is similar to looking up the name of a person in a telephone book to obtain their telephone number. However, if questions are asked for which there are no indexes then the whole of the file has to be read to obtain the information. With a random access file this is generally a much slower process than reading the whole of a sequential file. This is like trying to find the name of the person with a particular telephone number in a conventional telephone book. Ultimately, the choice depends upon the size of the file, the frequency with which it will be accessed, and the variety of criteria that will be used to retrieve data. Obviously, one does not want to completely read a large file very frequently! As there is an almost infinite number of criteria that might be used to retrieve data from most petrological databases, such as rock name, locality, field occurrence, mineral assemblage, age, ratios and functions of oxides, trace elements, normative values etc., it would seem logical to use sequential files wherever possible.

This conclusion is reinforced by over 10 years of practical experience in the use of petrological databases, as it has been found common practice to want to process the data many times over, as results from each run invariably suggest further refinements to be tried. Under such conditions the efficient way to proceed is to read the master database once only and to copy the required information onto a second file. This smaller file can then be read many times at a fraction of the cost of processing the entire database. Furthermore, if the system ensures that all the files it writes are in the same format as the files it reads, the same programs can process all the files.

Another useful attribute of any data processing system is the ability to work in either interactive mode, when the program runs under control from the terminal, or in batch mode, when the job is sent to a central site to await processing. For example, the typical *modus operandi* of a CLAIR job is to prepare any necessary Fortran and input data on a terminal; check to see that it works as desired on a small sub-set of data; and then batch the job to run with the required database and to obtain a hardcopy of the output.

Two systems, CLAIR and KEYBAM will now be briefly discussed as they were both designed to process large petrological databases.

12.3.1 The CLAIR data processing system

The current CLAIR data processing system, which is described in detail by Le Maitre & Ferguson (1978), has been used extensively by staff and students at the University of Melbourne over the last 7 years on both a CDC Cyber 73 and a Vax 11/780. In the last few years it has also been installed at over 10 other sites on a variety of other computers. Although the system was originally intended to process petrological data, there is nothing inherently petrological in its design, so that it has applications in many areas of data storage, retrieval and processing. It is, for example, just as capable of processing the CLAIR, PETROS, or IGBA databases, as it is of being used as a small bibliographic retrieval system. Some of its more useful general features are:-

- it is an open-ended series of 16 programs and over 250 subroutines written in ANSI Fortran IV, to store, retrieve and process both numeric and alphanumeric data.
- versions are avaliable for computers with 60-, 48-, 36- and 32-bit words and other versions can be produced on request.
- it works in both batch and interactive mode, when it sends prompts to the terminal requesting data input, and adjusts the width of the output to suit the width of the terminal.
- in standard versions each *data item* (a set of observations) can contain up to 180 floating point numbers and 800 alphanumeric characters, but versions containing any other number can easily be created.
- to save storage space, options exist to suppress zero numbers from the data items and to pack two floating point numbers into one computer word.
- all data files are sequential and written in an identical format in binary mode to speed up data transfer rates.
- data items may be organized into as many *CLAIR groups* as the user finds convenient, which is a feature designed to facilitate data retrieval, as all the programs access the data by group number.
- during execution of each program, the data items are read into unlabelled Common one at a time, where they may be modified or deleted by using one of the *optional user subroutines* called ADJUST, thus achieving complete flexibilty of selection and manipulation of the data.
- all the important parameters within the programs and subroutines are stored in unlabelled Common and labelled Common blocks, so that they may be monitored or changed by the optional user subroutines, if required.
- abundant error trapping exists and if the job is running interactively and the

error is recoverable, the data is requested again, otherwise the job can be made to terminate with its own dump of Common.
- most jobs run in under 70K octal on a CDC Cyber 73, unless the optional user subroutines are extremely long or use large arrays.

Among the general programs and subroutines are ones that:-
- create, modify, sort and copy data
- change group structures
- perform principal components analysis, multiple discriminant analysis, cluster analysis, multiple regression analysis and petrological mixing
- produce histogram data and plots
- print data items in a variety of tabular forms
- produce a variety of triangular, quadrilateral, rectangular or circular plots on a line-printer using either incremental plotting, with the option of simulated half-tone printing, or normal character plotting
- print arrays of characters across the page without truncation of words
- extract and insert characters into words
- move character strings from one word to another
- identify whether a character is alphabetic, numeric or neither
- search for the occurrence of character strings in arrays

These last four sets of subroutines are designed to facilitate the decoding of alphanumeric data for the purposes of data retrieval.

Among the specific petrological subroutines is one that calculates molecular weights given a chemical formula as characters, which was used extensively in the preparation of Tables A9 to A14 in the Appendix. Others calculate and print CIPW norms and cations on a basis of a specified number of oxygens. Whenever atomic or molecular weights are required by any of these subroutines, they are calculated once from the data of Table A8 in the Appendix which is stored in a common block.

12.3.2 The KEYBAM package

This was designed to process the PETROS database and has been described in detail by Barr et al. (1977). Many of its attributes are similar to those of the CLAIR data processing system, but it is not so flexible (F. Mutschler, per. comm.). It does, however, have the capability of being interfaced to the SPSS package (Nie et al. 1975).

12.4 CONCLUSIONS

Any readers that have persevered this far into the book should have a reasonably good theoretical understanding of the types of data processing that can be useful in petrology. All that remains is for them to get their databases organized, acquire a data processing package, and learn the practical aspects of the interpretation of petrological data. After all, there is nothing like experience.

REFERENCES

Abbey, S., 1977. "Standard samples": how standard are they? Geostand. Newsl., 1: 39-45.

Ahrens, L.H., 1954. The lognormal distribution of the elements. Geochim. Cosmochim. Acta, 5: 49-73.

Ahrens, L.H., 1977. A story of two rocks. Geostand. Newsl., 1: 157-161.

Aitchison, J., 1981. A new approach to null correlations of proportions. J. Int. Assoc. Math. Geol., 13: 175-189.

Albarède, F. and Provost, A., 1977. Petrological and geochemical mass-balance equations: an algorithm for least-squares fitting and general error analysis. Comput. Geosci., 3: 309-326.

Anderson, T.W., 1963. Asymptotic theory for principal components analysis. Ann. Math. Stat., 34: 122-148.

Baker, I., 1969. Petrology of the volcanic rocks of Saint Helena Island, South Atlantic. Bull. Geol. Soc. Am., 80: 1283-1310.

Baker, P.E., 1968. Petrology of Mt. Misery Volcano, St. Kitts, W. Indies. Lithos, 1: 124-150.

Baker, P.E., Gass, I., Harris, P.G. and Le Maitre, R.W., 1964. The volcanological report of the Royal Society Expedition to Tristan da Cunha. Phil. Trans. Roy. Soc. London. Ser. A, 256: 439-578.

Banks, R., 1979. The use of linear programming in the analysis of petrological mixing models. Contrib. Mineral. Petrol., 70: 237-244.

Barr, D.L., Mutschler, F.E. and Lavin, O.P., 1977. KEYBAM a system of interactive computer programs for use with the PETROS petrochemical data bank. Comput. Geosci., 3: 489-496.

Beswick, A.E. and Soucie, G., 1978. A correction procedure for metasomatism in an Archean Greenstone belt. Precambrian Res., 6: 235-248.

Binns, R.A., 1964. Zones of progressive regional metamorphism in the Willyama complex, Broken Hill district, New South Wales. J. Geol. Soc. Aust., 11: 283-330.

Binns, R.A., 1965a. The mineralogy of metamorphosed basic rocks from the Willyama Complex, Broken Hill district, New South Wales. Part I, Hornblendes. Mineral. Mag., 35: 306-326.

Binns, R.A., 1965b. The mineralogy of metamorphosed basic rocks from the Willyama Complex, Broken Hill district, New South Wales. Part II, Pyroxenes, garnets, plagioclases and opaque oxides. Mineral. Mag., 35: 561-587.

Bjerhammar, A., 1973. Theory of errors and generalised matrix inversions. Elsevier, Amsterdam, 420 pp.

Blackith, R.E. and Reyment, R.A., 1971. Multivariate Morphometrics. Academic Press, London, 412 pp.

Bowen, N.L., 1928. The evolution of the igneous rocks. Princeton Univ. Press, Princeton, 332 pp.

Box, G.E.P. and Cox, D.R., 1964. An analysis of transformations. J. Roy. Stat. Soc. Ser. B., 26: 211-243.

Brooks, C., Hart, S.R. and Wendt, I., 1972. Realistic use of two-error regression treatments as applied to Rubidium-strontium data. Rev. Geophys. Space Phy., 10: 551-577.

Brooks, C., Wendt, I. and Harre, W., 1968. A two-error regression treatment and its application to Rb-Sr and initial Sr^{87}/Sr^{86} ratios of younger Varisean granitic rocks from the Schwarzwald Massif, Southwest Germany. J. Geophys. Res., 73: 6071-6084.

Bryan, W.B., 1969. Materials balance in igneous rock suites. Annu. Rep. Dir. Geophys. Lab. Carnegie Inst. Wash., 67: 241-243.

Bryan, W.B., Finger, L.W. and Chayes, F., 1969a. A least-squares approximation for estimating the composition of a mixture. Annu. Rep. Dir. Geophys. Lab. Carnegie Inst. Wash., 67: 243-244.

Bryan, W.B., Finger, L.W. and Chayes, F., 1969b. Estimating proportions in petrographic mixing equations by least-squares approximation. Science, 163: 926-927.

Burk, C.F. Jr., 1981. International review of Geoscience source databases (Summary and lists). Aust. Min. Foundation, Adelaide, Seminar Project No. 154/81, 20 pp.

Butler, J.C., 1975. Occurrence of negative open variances in ternary systems. J. Int. Assoc. Math. Geol., 7: 31-45.

Butler, J.C., 1976. Principal components analysis using the hypothetical closed array. J. Int. Assoc. Math. Geol., 8: 25-36.

Butler, J.C., 1979a. Trends in ternary petrologic variation diagrams - fact or fantasy? Am. Mineral., 64: 1115-1121.

Butler, J.C., 1979b. Numerical consequences of changing the units in which chemical analyses of igneous rocks are analysed. Lithos, 12: 33-39.

Butler, J.C., 1980. Numerical consequences of computing structural formulas. Lithos, 13: 55-59.

Cacoullos, T. and Styan, P.H., 1973. A bibliography of discriminant analysis. In: Cacoullos, T. (Editor), Discriminant analysis and applications. Academic Press, New York, 375-434.

Carmichael, D.M., 1970. Intersecting isograds in the Whetstone Lake area, Ontario. J. Petrol., 11: 147-181.

Cattell, R.B., 1965. Factor analysis: an intoduction to essentials. Biometrics, 21: 190-215.

Chayes, F., 1956. Petrographic modal analysis. J. Wiley & Sons, New York, 113 pp.

Chayes, F., 1960. On correlation between variables of constant sum. J. Geophys. Res., 65: 4185-4193.

Chayes, F., 1962. Numerical correlation and petrographic variation. J. Geol., 70: 440-452.

Chayes, F., 1964. Variance-covariance relations in some published Harker diagrams of volcanic suites. J. Petrol., 5: 219-237.

Chayes, F., 1967. On the graphical appraisal of the strength of association in petrographic variation diagrams. In: Abelson, P.H. (Editor), Researches in Geochemistry, Vol. 2. J. Wiley & Sons, New York, 322-339.

Chayes, F., 1968. A least square approximation for estimating the amounts of petrographic partition products. Mineral. Petrogr. Acta, 14: 111-114.

Chayes, F., 1970a. Another last look at G1-W1. J. Int. Assoc. Math. Geol., 2: 207-209.

Chayes, F., 1970b. Effect of a single nonzero open covariance on the simple closure test. In: Merriam, D.F. (Editor), Geostatistics. Plenum Press, New York, 11-22.

Chayes, F., 1970c. On estimating the magnitude of the Hidden zone and the composition of the residual liquids of the Skaergaard Layered Series. J. Petrol., 11: 1-14.

Chayes, F., 1970d. On deciding whether trend surfaces of progressively higher order are significant. Bull. Geol. Soc. Am., 81: 1273-1278.

Chayes, F., 1971. Ratio Correlation. Univ. Chicago Press, Chicago, 98 pp.

Chayes, F., 1972. Statistical petrography. Annu. Rep. Dir. Geophys. Lab. Carnegie Inst. Wash., 71: 489-495.

Chayes, F., 1976. Asymmetry in the distributions of SiO_2, Al_2O_3, CaO and Fe in the products of Cenozoic volcanism. Annu. Rep. Dir. Geophys. Lab. Carnegie Inst. Wash., 75: 780-781.

Chayes, F., 1979. A world data base for igneous petrology. Annu. Rep. Dir. Geophys. Lab. Carnegie Inst. Wash., 78: 484-485.

Chayes, F. and Kruskal, W., 1966. An approximate statistical test for correlations between proportions. J. Geol., 74: 692-702.

Chayes, F. and Le Maitre, R.W., 1972a. The number of published analyses of igneous rocks. Annu. Rep. Dir. Geophys. Lab. Carnegie Inst. Wash., 71: 493-495.

Chayes, F. and Le Maitre, R.W., 1972b. Published analyses of igneous rocks. Nature, London, 236: 449-450.

Clanton, U.S. and Fletcher, C.R., 1976. Sample size and sampling errors as the source of dispersion in chemical analyses. Proc. 7th Lunar Sci. Conf., 1413-1428.

Clark, A.L., 1976. Resource data bases - resource assessment. Comput. Geosci., 2: 309-311.

Coats, R.R., 1952. Magmatic differentiation in Tertiary and Quaternary volcanic rocks from Adak and Kanaga Islands, Aleutian Islands, Alaska. Bull. Geol. Soc. Am., 63: 485-514.

Cooley, W.W. and Lohnes, P.R., 1971. Multivariate data analysis. J. Wiley & Sons, New York, 364 pp.

Cox, K.G. and Jamieson, B.G., 1974. The olivine-rich lavas of Nuanetsi: a study of polybaric magmatic evolution. J. Petrol., 15: 269-301.

Cox, K.G., Bell, J.D. and Pankhurst, R.J., 1979. The interpretation of igneous rocks. George Allen & Unwin, London, 450 pp.

Creasy, M.A., 1956. Confidence limits for the gradient in the linear functional relationship. J. Roy. Stat. Soc. Ser. B, 18: 65-69.

Cross, W., Iddings, J.P., Pirsson, L.V. and Washington, H.S., 1903. Quantitative classification of igneous rocks. Univ. Chicago Press, Chicago, 286 pp.

Darroch, J.N., 1969. Null correlation for proportions. J. Int. Assoc. Math. Geol., 1: 221-227.

Darroch, J.N. and Ratcliff, D., 1970. Null correlation for proportions II. J. Int. Assoc. Math. Geol., 2: 307-312.

Darroch, J.N. and Ratcliff, D., 1978. No-association of proportions. J. Int. Assoc. Math. Geol., 10: 361-368.

Das Gupta, S., 1973. Theories and methods in classification: a review. In: Cacoullos, T. (Editor), Discriminant analysis and applications. Academic Press, New York, 77-137.

Davis, J.C., 1973. Statistics and data analysis in geology. J. Wiley & Sons, New York, 550 pp.

Deer, W.A., Howie, R.A. and Zussman, J., 1966. An introduction to the rock forming minerals. Longmans, London, 528 pp.

Dixon, W.J. and Massey, F.J., 1969. Introduction to statistical analysis. McGraw-Hill, New York, 638 pp.

Draper, N.R. and Smith, H., 1966. Applied regression analysis. J. Wiley & Sons, New York, 407 pp.

Engels, J.C. and Ingamells, C.O., 1977. Geostandards - a new approach to their production and use. Geostand. Newsl., 1: 51-60.

Eskola, P., 1915. On the relations between the chemical and mineralogical compositions in the metamorphic rocks of the Orijarvi Region. Bull. Comm. Geol. Fin., No. 44, 1-145.

Ewart, A. and Le Maitre, R.W., 1980. Some regional compositional differences within Teriary - Recent orogenic magmas. Chem. Geol., 30: 257-283.

Fenner, C.N., 1926. The Katmai magmatic province. J. Geol., 34: 673-772.

Fenner, C.N., 1950. The chemical kinetics of the Katmai eruption. Part I. Am. J. Sci., 248: 593-627.

Fitzgerald, J.D. and Mackinnon, I.D.R., 1977. PETPAK - a computing package for the petrologist. Comput. Geosci., 3: 637-638.

Flanagan, F.J., 1976a. Introduction. In: Flanagan F.J. (Editor), Description and analyses of eight new USGS rock standards. Prof. Pap. U.S. Geol. Surv., 840: 1-5.

Flanagan, F.J., 1976b. 1972 compilation of data on USGS standards. In: Flanagan F.J. (Editor), Description and analyses of eight new USGS rock standards. Prof. Pap. U.S. Geol. Surv., 840: 131-183.

Flanagan, F.J., 1976c. G-1 et W-1: Requiescant in Pace! In: Flanagan F.J. (Editor), Description and analyses of eight new USGS rock standards. Prof. Pap. U.S. Geol. Surv., 840: 189-192.

Francis, I., 1974. Factor analysis: fact or fabrication. Math. Chronicle, 3: 9-44.

Frey, F.A., Green, D.H. and Roy, S.D., 1978. Integrated models of basalt petrogenesis: a study of quartz tholeiites to olivine melelitites from South Eastern Australia utilizing geochemical and experimental petrological data. J. Petrol., 19: 463-513.

Full, W.E., Ehrlich, R. and Klovan, J.E., 1981. EXTENDED QMODEL - objective definition of external end members in the analysis of mixtures. J. Int. Assoc. Math. Geol., : 331-344.

Gill, D., Beylin, J., Boehm, S., Frenkel, Y. and Rosenthal, E., 1977. Design of geological data systems for developing nations. J. Int. Assoc. Math. Geol., 9: 145-157.

Gilliland, J.A. and Grove, G., 1973. Some principles of data storage and information retrieval and their implications for information exchange. J. Int. Assoc. Math. Geol., 5: 1-10.

Gnanadesikan, R., 1977. Methods for statistical data analysis of multivariate observations. J. Wiley & Sons, New York, 311 pp.

Goubin, N., 1978. Some examples of management and processing of geological and geochemical data. Comput. Geosci., 4: 37-52.

Gower, J.C., 1967. A comparison of some methods of cluster analysis. Biometrics, 23: 623-637.

Grandclaude, Ph., 1976a. Design and use of a geochemical data bank. Comput. Geosci., 2: 163-170.

Grandclaude, Ph., 1976b. Nature and structure of the information in the earth sciences. Some theoretical and practical (file and data bank) aspects. Sci. Terre Inf. Geol., 10: 3-40.

Gray, N.H., 1973. Estimation of parameters in petrologic materials balance equations. J. Int. Assoc. Math. Geol., 5: 225-236.

Greenwood, H.J., 1967. The N-dimensional tie line problem. Geochim. Cosmochim. Acta, 31: 465-490.

Greenwood, H.J., 1968. Matrix methods and the phase rule in petrology. 23rd Int. Geol. Congr. (Prague), 6: 267-279.

Gy, P.M., 1979. Sampling of particulate materials, theory and practice. Elsevier, Amsterdam, 431 pp.

Harbaugh, J.W. and Bonham-Carter, J., 1970. Computer Simulation in Geology. J. Wiley & Sons, New York, 575 pp.

Harrison, R.K. and Sabine, P.A., 1970. A petrological-mineralogical code for computer use. Inst. Geol. Sci., Report 70/6, 134 pp.

Hartigan, J.A., 1975. Clustering algorithms. J. Wiley & Sons, New York, 351 pp.

Hawkins, D.B., 1974. Statistical analyses of the zeolites clinoptilolite and heulandite. Contrib. Mineral. Petrol., 45: 27-36.

Hey, M.H., 1969. The determination of multiple correlations between several variables with especial reference to the correlation of physical properties and chemical composition. Mineral. Mag., 37: 83-89.

Horder, M.F., 1981. The use of databanks and databases within the Institute of Geological Sciences. J. Geol. Soc., 138: 575-582.

Howarth, R.J. and Earle, S.A.M., 1979. Application of a generalised power transformation to geochemical data. J. Int. Assoc. Math. Geol., 11: 45-62.

Hubaux, A., 1969. Archival files of geological data. J. Int. Assoc. Math. Geol., 1: 41-52.

Hutchison, C.S., 1975. Correlation of Indonesian active volcanic geochemistry with Benioff zone depth. Geol. Mijnbouw, 54: 157-168.

Imbrie, J., 1963. Factor and vector analysis programs for analyzing geologic data. Office Naval Res., Geog. Branch, Tech. Rep. 6 (ONR Task No. 389-135), 83 pp.

Imbrie, J. and Purdy, E.G., 1962. Classification of modern Bahamian carbonate sediments. In: Classification of carbonate rocks - a symposium. Am. Assoc. Petroleum Geol., Mem. 1, 253-272.

Ingamells, C.O., 1974. Control of geochemical error through sampling and subsampling diagrams. Geochim. Cosmochim. Acta, 38: 1225-1237.

Ingamells, C.O., Engels, J.C. and Switzer, P., 1972. Effect of laboratory sampling error in geochemistry and geochronology. Proc. 24th Int. Geol. Congr. (Montreal), 10: 405-415.

Jeffery, K.G. and Gill, E.M., 1976. The design philosphy of the G-EXEC system. Comput. Geosci., 2: 345-349.

Jeffery, K.G. and Gill, E.M., 1977. The use of G-EXEC for resource analysis. J. Int. Assoc. Math. Geol., 9: 265-272.

Johannsen, A., 1950. A descriptive petrography of the igneous rocks. Univ. Chicago Press, Chicago, 1: 318 pp.

Jones, T.A., 1972. Multiple regression with correlated independant variables. J. Int. Assoc. Math. Geol., 4: 203-218.

Jones, T.A., 1979. Fitting straight lines when both variables are subject to error. I. Maximum likelihood and least-squares estimation. J. Int. Assoc. Math. Geol., 11: 1-25.

Jöreskog, K.G., Klovan, J.E. and Reyment, R.A., 1976. Geological factor analysis. Elsevier, Amsterdam, 178 pp.

Kelsey, C.H., 1965. Calculation of the C.I.P.W. norm. Mineral. Mag., 34: 276-282.

Kendall, M.G., 1943. The advanced theory of statistics. Vol. I. Griffin & Co., London, 457 pp.

Kendall, M.G., 1973. The basic problems of cluster analysis. In: Cacoullos, T. (Editor), Discriminant analysis and applications. Academic Press, New York, 179-191.

Kim, C., 1971. Introduction to linear programming. Holt, Rinehart & Winston, New York, 556 pp.

Kleeman, A.W., 1967. Sampling error in the chemical analysis of rocks. J. Geol. Soc. Aust., 14: 43-47.

Klovan, J.E., 1981. A generalization of extended Q-mode factor analysis to data matrices with variable row sums. J. Int. Assoc. Math. Geol., 13: 217-224.

Koch, G.S. and Link, R.F., 1970. Statistical analysis of geological data. J. Wiley & Sons, New York, 375 pp.

Koch, G.S. and Link, R.F., 1971. Statistical analysis of geological data. Vol II. J. Wiley & Sons, New York, 438 pp.

Kork, J.O., 1977. Examination of the Chayes-Kruskal procedure for testing correlations between proportions. J. Int. Assoc. Math. Geol., 9: 543-562.

Kremer, M., Lenci, M. and Lesage, M.T., 1976. SIGMA: a user-orientated file-processing system. Comput. Geosci., 1: 187-193.

Lawley, D.N. and Maxwell, A.E., 1971. Factor analysis as a statistical method. Butterworths, London, 153 pp.

Le Maitre, R.W., 1962. Petrology of volcanic rocks, Gough Island, S. Atlantic. Bull. Geol. Soc. Am., 73: 1309-1340.

Le Maitre, R.W., 1965. Comments on "A method of classifying analyses with any number of items" by J.B. Mertie, Jr. Am. Mineral., 50: 1131-1133.

Le Maitre, R.W., 1968. Chemical variation within and between volcanic rock series - a statistical approach. J. Petrol., 9: 220-252.

Le Maitre, R.W., 1973. Experiences with CLAIR: a computerised library of analysed igneous rocks. Chem. Geol., 12: 301-308.

Le Maitre, R.W., 1976a. Chemical variability of some common igneous rocks. J. Petrol., 17: 589-637.

Le Maitre, R.W., 1976b. Some problems of the projection of chemical data into mineralogical classifications. Contrib. Mineral. Petrol., 56: 181-189.

Le Maitre, R.W., 1976c. A new approach to the classification of igneous rocks using the basalt-andesite-dacite-rhyolite suite as an example. Contrib. Mineral. Petrol., 56: 191-203.

Le Maitre, R.W., 1978. The use of the CLAIR data system and the transferal of data files. Sci. Terre Inf. Geol. No. 11, 43-44.

Le Maitre, R.W., 1979. A new generalised petrological mixing model. Contrib. Mineral. Petrol., 71: 133-137.

Le Maitre, R.W., 1980. Numerical petrology. Trans. Leic. Lit. Phil. Soc. (for 1978), 72: 70-96.

Le Maitre, R.W., 1981. GENMIX - A generalised petrological mixing model program. Comput. Geosci., 7: 229-247.

Le Maitre, R.W. and Ferguson, A.K., 1978. The CLAIR data system. Comput. Geosci., 4: 65-76.

Lindley, D.V., 1947. Regression lines and the linear functional relationship. J. Roy. Stat. Soc. Supp., 9: 218-244.

Longely, J.W., 1967. An appraisal of least squares programs for the electronic computer from the point of view of the user. J. Am. Stat. Soc., 62: 819-841.

Loudon, T.V., 1969. A small geological data library. J. Int. Assoc. Math. Geol., 1: 155-170.

MacDonald, G.A., 1949. Hawaiian Petrographic Province. Bull. Geol. Soc. Am., 60: 1541-1596.

Mark, D.M. and Church, M., 1977. On the misuse of regression in Earth Science. J. Int. Assoc. Math. Geol., 9: 63-75.

Marriott, F.H.C., 1974. The interpretation of multiple observations. Academic Press, London, 117 pp.

McBirney, A.R. and Aoki, K., 1968. Petrology of the Island of Tahiti. Mem. Geol. Soc. Am., 116: 523-556.

McIntyre, G.A., Brooks, C., Compston, W. and Turek, A., 1966. The statistical assessment of Rb-Sr isochrons. J. Geophys. Res., 71: 5459-5468.

Miesch, A.T., 1967a. Theory of error in geochemical data. Prof. Pap. U.S. Geol. Surv., 574-A: 1-17.

Miesch, A.T., 1967b. Methods of computation for estimating geochemical abundances. Prof. Pap. U.S. Geol. Surv., 574-B: 1-15.

Miesch, A.T., 1969. The constant sum problem in geochemistry. In: Merriam, D.F. (Editor), Computer applications in the Earth Sciences. Plenum Press, New York, 161-176.

Miesch, A.T., 1976a. Geochemical survey of Missouri - methods of sampling, laboratory analysis, and statistical reduction of data. Prof. Pap. U.S. Geol. Surv., 954-A: 1-39.

Miesch, A.T., 1976b. Q-mode factor analysis of compositional data. Comput. Geosci., 1: 147-159.

Miesch, A.T., 1976c. Interactive computer program for petrologic modelling with extended Q-mode factor analysis. Comput. Geosci., 2: 439-492.

Miesch, A.T., 1976d. Q-mode factor analysis of geochemical and petrologic data matrices with constant row-sums. Prof. Pap. U.S. Geol. Surv., 574-G: 1-47.

Miesch, A.T., 1980. Scaling variables and interpretation of eigenvalues in principal components analysis of geological data. J. Int. Assoc. Math. Geol., 12: 523-538.

Miesch, A.T., 1981. Computer methods for geochemical and petrological mixing problems. In: Merriam, D.F. (Editor), Computer applications in the Earth Sciences. Plenum Press, New York, 243-265.

Miesch, A.T., Chao, E.C.T. and Cuttitta, F., 1966. Multivariate analysis of geochemical data on tektites. J. Geol., 74: 673-691.

Miller, R.L. and Kahn, J.S., 1962. Statistical analysis in the geological sciences. J. Wiley & Sons, New York, 483 pp.

Moore, F., 1979. Some statistical calculations concerning the determination of trace constituents. Geostand. Newsl., 3: 105-108.

Moran, P.A.P., 1971. Estimating structural and functional relationships. J. Multivar. Anal., 1: 232-255.

Morrison, D.F., 1967. Multivariate statistical methods. McGraw-Hill, New York, 338 pp.

Mutschler, F.E., Rougon, D.J. and Lavin, O.P., 1976. PETROS - a data bank of major-element chemical analyses of igneous rocks for research and teaching. Comput. Geosci., 2: 51-57.

Mutschler, F.E., Rougon, D.J., Lavin, O.P. and Hughes, R.D., 1978. PETROS a data bank of major-element chemical analyses of igneous rocks. U.S. Dept. Commerce, Nat. Oceanic Atmos. Admin., Envir. Data Serv., Pamphlet 1978(W), .

Mutschler, F.E., Wright, E.G., Ludington, S. and Abbott, J.T., 1981. Granite molybdenite systems. Econ. Geol., 76: 874-897.

Nie, N.H., Hull, C.H., Jenkins, J.G., Steinbrenner, K. and Bent, D.H., 1975. SPSS: Statistical package for the social sciences. McGraw-Hill, New York, 675 pp.

Ono, K., 1962. Chemical composition of volcanic rocks in Japan. Publ. Geol. Surv. Jap., 459 pp.

Paraskevopoulos, G.M., 1956. Über den Chemismus und die provinzialen Verhältnisse der tertiären und quartären Ergußgesteine des ägäischen Raumes und der benachbarten Gebiete. Tschermaks Mineral. Petrogr. Mitt., 6: 13-72.

Parker, R.J., 1981. GEOIC: an interactive terminal-based geochemical data processing system. Comput. Geosci., 7: 287-296.

Pearce, T.H., 1968. A contribution to the theory of variation diagrams. Contrib. Mineral. Petrol., 19: 142-157.

Pearce, T.H., 1970. Chemical variations in the Palisade sill. J. Petrol., 11: 15-32.

Perry, K., 1967. An application of linear algebra to petrologic problems: Part 1. Mineral classification. Geochim. Cosmochim. Acta, 31: 1043-1078.

Powers, H.A., 1955. Composition and origin of basaltic magma of the Hawaiian Islands. Geochim. Cosmochim. Acta, 7: 77-107.

Provost, A. and Allegre, C.J., 1979. Process identification and search for optimal differentiation parameters from major element data. General presentation with emphasis on fractional crystallisation process. Geochim. Cosmochim. Acta, 43: 487-501.

Rao, C.R. and Mitra, S.K., 1971. Generalized inverse of matrices and its application. J. Wiley & Sons, New York, 240 pp.

Reid, M.J., Gancarz, A.J. and Albee, A.L., 1973. Constrained least-squares analysis of petrologic problems with an application to lunar sample 12040. Earth Planet. Sci. Lett., 17: 433-445.

Ridley, K.J.D., Turek, A. and Riddle, C., 1976. The variability of chemical analyses as a function of sample heterogeneity, and the implications to the analyses of rock standards. Geochim. Cosmochim. Acta, 40: 1375-1379.

Robinson, S.C., 1970. A review of data processing in the Earth Sciences in Canada. J. Int. Assoc. Math. Geol., 2: 377-397.

Rubeska, I., 1977. The state of art of trace element analysis of geological samples as derived from results on standard rock samples. Geostand. Newsl., 1: 15-20.

Russell, B.G., Goudvis, R.G., Domel, G. and Levin, J., 1972. Preliminary report on the analysis of the six NIMROC geochemical standard samples. Rep. Nat. Inst. Metal. Johannesburg, No.1351, 1-74.

Saha, A.K., Bhattacharyya, C. and Lakshmipathy, S., 1974. Some problems of interpreting the correlations between the modal variables in granitic rocks. J. Int. Assoc. Math. Geol., 6: 245-258.

Saxena, S.K., 1969a. A statistical approach to the study of phase equilibria in multicomponent systems. Lithos, 3: 25-36.

Saxena, S.K., 1969b. Silicate solid solutions and geothermometry. 4. Statistical study of chemical data on garnets and clinopyroxenes. Contrib. Mineral. Petrol., 23: 140-156.

Saxena, S.K., 1969c. Silicate solid solutions and geothermometry. 3. Distribution of Fe and Mg between coexisting garnet and biotite. Contrib. Mineral. Petrol., 22: 259-267.

Saxena, S.K. and Ekström, T.K., 1970. Statistical chemistry of calcic amphiboles. Contrib. Mineral. Petrol., 26: 276-284.

Saxena, S.K. and Walter, L.S., 1974. A statistical-chemical and thermodynamic approach to the study of lunar mineralogy. Geochim. Cosmochim. Acta, 38: 79-95.

Saxena, S.K., Benimoff, A. and Pingitore, N.E., 1977. Moon - an albite depleted giant Skaergaard? Contrib. Mineral. Petrol., 60: 77-90.

Shaw, D., 1969. Evaluation of data. In: Wedepohl, K.H. (Editor), Handbook of geochemistry. Vol.1. Springer-Verlag, Berlin, 324-375.

Simonen, A., 1948. On the petrology of the Aulanko area in southwestern Finland. Bull. Comm. Geol. Fin., 25: 1-66.

Skala, W., 1977. A mathematical model to investigate distortions of correlation coefficients in closed arrays. J. Int. Assoc. Math. Geol., 9: 519-528.

Skala, W., 1979. Some effects of the constant-sum problem in geochemistry. Chem. Geol., 27: 1-9.

Snedecor, G.W. and Cochran, W.G., 1967. Statistical methods. Iowa State Univ. Press, Iowa, 593 pp.

Snow, J.W., 1975. Association of proportions. J. Int. Assoc. Math. Geol., 7: 63-73.

Sommerville, D.M.Y., 1929. An introduction to the geometry of N dimensions. Metheun, London, 196 pp.

Spearman, C., 1904. General intellegence objectively determined and measured. Am. J. Psychol., 15: 201-293.

Steele, T.W. and Hansen, R.G., 1979. Major element data (1966-1978) for the six "NIMROC" reference standards. Geostand. Newsl., 3: 135-172.

Stevens, R.E. and Niles, W.W., 1960. Chemical analyses of the granite and the diabase. Bull. U.S. Geol. Surv., 1113: 3-43.

Stormer, J.C. and Nicholls, J., 1978. XLFRAC: a program for the interactive testing of magmatic differentiation models. Comput. Geosci., 4: 143-159.

Streckeisen, A., 1978. Classification and nomenclature of volcanic rocks, lamprophyres, carbonatites and melilitic rocks. Neues Jahrb. Mineral. Abh., 134: 1-14.

Streckeisen, A. and Le Maitre, R.W., 1979. A chemical approximation to the modal QAPF classification of the igneous rocks. Neues Jahrb. Mineral. Abh., 136: 169-206.

Stuckless, J.S. and Miesch, A.T., 1981. Petrogenetic modeling of a potential uranium source rock, Granite Mountains, Wyoming. Prof. Pap. U.S. Geol. Surv., 1225: 1-34.

Tatsuoka, M.M., 1971. Multivariate Analysis. J. Wiley & Sons, New York, 310 pp.

Temple, J.T., 1978. The use of factor analysis in geology. J. Int. Assoc. Math. Geol., 10: 379-387.

Thompson, J.B., 1957. The graphical analysis of mineral assemblages in pelitic schists. Am. Mineral., 42: 842-858.

Thornton, C.P. and Tuttle, O.F., 1960. Chemistry of igneous rocks; part I. Differentiation index. Am. J. Sci., 258: 664-684.

Thurstone, L.L, 1947. Multiple factor analysis. Univ. Chicago Press, Chicago, 535 pp.

Till, R., 1974. Statistical methods for the earth scientist. MacMillan, London, 154 pp.

Till, R., 1977. The HARDROCK package, a series of FORTRAN IV computer programs for performing and plotting petrochemical calculations. Comput. Geosci., 3: 185-243.

Till, R. and Colley, H., 1973. Thoughts on the use of principal component analysis in petrogenetic problems. J. Int. Assoc. Math. Geol., 5: 341-350.

Tröger, W.E., 1969. Spezielle Petrographie der Eruptivgesteine. Verlag. Deuts. Min. Gesell. Bonn, 360 + 90 pp.

Trochimczyk, J. and Chayes, F., 1977. Sampling variation of principal components. J. Int. Assoc. Math. Geol., 9: 497-506.

Trochimczyk, J. and Chayes, F., 1978. Some properties of principal component scores. J. Int. Assoc. Math. Geol., 10: 43-52.

Trustrum, K., 1971. Linear programming. Routledge & Kegan Paul, London, 88 pp.

Turner, F.J., 1968. Metamorphic petrology. McGraw-Hill, New York, 403 pp.

van Trump, G. and Miesch, A.T., 1977. The U.S. Geological Survey RASS-STATPAC system for managment and statistical reduction of geochemical data. Comput. Geosci., 3: 475-488.

Vistelius, A.B., 1970. Statistical model of silicate analysis and results of investigation of G-1 and W-1 samples. J. Int. Assoc. Math. Geol., 2: 1-14.

Vistelius, A.B. and Sarmanov, O.V., 1961. On the correlation between percentage values: major component correlation in ferromagnesium micas. J. Geol., 69: 145-153.

Vlodavetz, V.I. and Piip, B.I., 1959. Catalogue of the active volcanoes of the World. Part 8. Kamchatka and Continental areas of Asia. Int. Assoc. Volcan., VIII: 1-110.

Walpole, R.E. and Myers, R.H., 1978. Probability and statistics for engineers and scientists. MacMillan, New York, 580 pp.

Washington, H.S., 1917. Chemical analyses of igneous rocks. Prof. Pap. U.S. Geol. Surv., 99: 1-1201.

Weast, R.C. (Editor), 1974. Handbook of Chemistry and Physics. CRC Press Inc., Cleveland, .

Westerveld, J., 1952. Quaternary volcanism on Sumatra. Bull. Geol. Soc. Am., 63: 561-594.

Williams, E.J., 1959. Regression analysis. J. Wiley & Sons, New York, 211 pp.

Williams, E.J., 1980. Generalized matrix inverses with specified properties. Unpubl. Res. Rep. No.4, Dept. Stat. Univ. Melb., 12 pp.

Williams, H., 1942. The geology of Crater Lake National Park, Oregon. Publ. Carnegie Inst. Wash., 540: 1-157.

Wishart, D., 1969. Fortran II programs for 8 methods of cluster analysis (Clustan I). Computer Contrib. 38, Kansas State Geol. Surv., 112 pp.

Wonnocott, T.H. and Wonnacott, R.J., 1977. Introductory statistics. J. Wiley & Sons, New York, 650 pp.

Wright, T.L., 1974. Presentation and interpretation of chemical data for igneous rocks. Contrib. Mineral. Petrol., 48: 233-248.

Wright, T.L. and Doherty, P.C., 1970. A linear programming and least squares computer method for solving petrological mixing problems. Bull. Geol. Soc. Am., 81: 1995-2008.

York, D., 1969. Least squares fitting of a straight line with correlated errors. Earth Planet. Sci. Lett., 5: 320-324.

Zodrow, E.L., 1976. Empirical behaviour of Chayes' null model. J. Int. Assoc. Math. Geol., 8: 37-42.

APPENDIX

STATISTICAL, CHEMICAL AND MINERALOGICAL TABLES

TABLE A1

5,000 Random digits, calculated using IMSL subroutine GGUBS

	0	1	2	3	4	5	6	7	8	9
1	05053	65112	62946	30068	02011	31894	91158	95083	97608	85622
2	17823	98797	80188	91038	06393	52612	95124	58796	76210	19029
3	46418	86349	49664	19555	02093	52818	68816	69701	04677	97708
4	55538	55846	97643	81739	56198	30634	14718	51903	60280	54528
5	24842	44129	95274	56492	62575	80139	36564	85371	52663	59853
6	00278	01140	59191	55596	43623	43403	31379	86721	98808	26635
7	30066	17834	13609	47983	76876	37021	14737	63940	01176	35285
8	29828	80152	91275	01691	34829	44290	83300	09998	39200	28565
9	65954	37427	12236	68965	67423	84039	19685	02358	98144	46223
10	25642	06100	77913	41935	35011	19250	46399	22032	99010	47678
11	34688	99959	43849	21696	22273	76939	40425	08439	22040	95940
12	03622	34488	90642	56303	95159	57580	19125	65302	71519	07965
13	54827	21640	34915	54726	37342	30089	74188	62477	96612	74893
14	88413	12272	20832	30176	06500	32159	01110	90262	89782	33015
15	30980	28073	62849	25613	22188	38875	43234	32891	30183	77708
16	02333	40494	22112	15009	68843	63946	99676	56098	67770	82083
17	62334	06732	09950	29658	55309	01077	52675	83552	49775	86751
18	93351	89849	81673	96885	03530	92014	52170	93818	54176	77552
19	60374	02772	00511	14046	54580	91574	98161	75468	47900	51220
20	45538	63193	52909	97203	62360	00977	14984	30344	90593	48199
21	21596	39024	00640	30154	83224	99793	31179	13288	72247	23947
22	55260	36568	65808	03480	36699	47741	22233	69928	50608	69123
23	05131	04890	49125	23121	75053	54113	92833	43293	78155	92947
24	79551	33298	31987	22130	59830	32735	45154	08558	32928	47357
25	96677	18687	93264	00860	67039	12153	53948	71384	77543	60586
26	93251	99675	03737	63816	52778	65239	83637	26946	95505	16938
27	26761	64155	99815	32504	91928	56722	98018	15979	98069	30304
28	41203	08743	50903	92743	67739	53567	30239	54129	86496	36127
29	13635	13361	70627	02090	11300	28681	43783	59401	93375	92366
30	95087	29560	90159	60646	10800	58330	83162	04798	24391	93078
31	60729	85624	09584	02990	62841	98355	46202	93552	06082	18915
32	59911	65213	01745	75962	71432	56371	75032	56689	29293	29073
33	79470	85541	68902	42028	55695	14894	59831	62438	26699	51203
34	58667	69723	01369	76249	26139	73865	19552	29712	67696	70979
35	60641	92974	50456	18239	29589	80685	55861	47956	78092	96114
36	49335	44099	12749	13848	32152	70067	27650	67115	08171	09873
37	30071	53048	56907	00266	50705	60717	22340	67992	31449	90063
38	13559	94713	54841	09058	33146	87580	86591	20880	55719	84117
39	41285	46744	58352	01602	10646	43729	42623	99555	29152	60042
40	43747	49163	27686	86884	24087	27957	18482	13985	39688	68454
41	07324	62643	18173	26589	91261	78987	31365	07731	03900	95934
42	30039	84980	02175	72582	14049	73437	55141	67435	52862	41861
43	38234	05021	28061	60282	63819	31522	55379	88788	25070	28653
44	85051	66757	71593	06945	48424	81381	28277	96174	48618	04238
45	26061	40377	18653	17793	62239	53958	62440	89952	15769	82723
46	21858	30737	52403	17743	06489	82063	09890	07142	94457	02181
47	89715	59152	45352	69124	45091	00997	25087	72657	98420	59028
48	41142	47690	00683	27862	81314	09119	86701	71189	66035	78421
49	24564	59513	16501	54317	78827	79933	39452	81244	74925	17158
50	52756	68216	38415	96897	92832	39590	53572	20300	78445	62211

	0	1	2	3	4	5	6	7	8	9
51	30979	67666	10473	29929	22655	98540	12986	62729	59613	80380
52	88039	82578	84264	31539	30320	42405	62592	73776	20498	03811
53	57936	17187	57934	78038	00261	19736	37826	45780	81252	42276
54	26368	16641	79834	80734	02439	72293	64136	80488	34129	93991
55	24391	26667	45431	32354	15557	94202	65037	16894	63323	77918
56	67129	32714	17070	32393	34994	18342	48533	27034	50244	57063
57	33389	35514	27677	25625	55173	22427	60589	87196	38276	97648
58	57804	05332	90232	38338	23096	22677	03819	33218	00194	69702
59	42321	09860	32011	57560	60778	15237	67811	78180	43632	33767
60	98813	39803	93333	88170	92437	85419	08801	07758	43319	84618
61	71873	76823	08275	07734	78956	27810	25334	94957	57908	55163
62	96946	93610	80811	04462	10267	83945	63408	38321	94416	97041
63	09912	76570	63715	91390	41330	60693	34135	44500	01223	22213
64	19895	34411	89249	44071	84163	38384	04011	15356	71871	91841
65	13764	93547	72413	54224	10220	58960	18570	98546	38664	34749
66	87128	03882	58219	14362	94394	90237	85181	74026	08638	13151
67	87681	99586	33273	14657	65078	59723	76667	16660	92921	34295
68	47174	94279	76997	14194	63548	61929	14187	49456	82891	14825
69	98366	92556	58234	42498	38207	79733	09359	92703	03457	76950
70	02318	50371	35552	61414	25048	60989	06812	59762	54736	12090
71	54468	09312	66074	89897	79411	81600	94310	39138	62253	15089
72	96379	45660	57968	35303	67700	51385	25566	12474	01157	65233
73	98151	75400	56525	66942	65408	33211	52132	99246	06760	88638
74	73382	99011	04306	33836	13006	65719	90781	35868	33342	89710
75	85233	10312	50447	42578	01396	54941	85228	52203	98389	41702
76	86099	30070	02526	94378	43825	25879	54784	75990	59083	77143
77	79914	43747	85619	25186	26951	09043	05508	14524	53920	46296
78	48109	31953	56571	11237	00268	28201	90573	63648	21112	47688
79	19285	03527	30888	42967	77743	67119	72943	11079	34167	13303
80	15676	57874	02825	09718	75982	26177	12147	01795	75627	20222
81	68356	51948	32642	04332	73257	45022	99448	20253	35559	41881
82	18355	26822	25684	20486	99092	68040	98122	24549	62840	83537
83	20210	51922	92197	58389	55919	76292	37734	41997	87169	01591
84	24981	77490	05848	19838	84395	14987	55146	16767	26688	81508
85	83334	14519	29035	21483	27623	06446	07758	85956	56893	94667
86	64860	79843	58854	78861	21769	24714	76632	59130	51346	49700
87	06688	56070	45663	88983	61580	83926	45741	58969	84588	19267
88	90538	86490	63130	94433	70313	01578	77126	38701	49931	11765
89	86429	88404	63306	63428	64368	63412	45692	55742	11788	81664
90	46623	59085	72981	65576	34958	89676	47544	34737	84018	32065
91	18728	03753	45503	90800	77917	41021	08720	64827	26450	04960
92	54172	09468	11970	37105	15017	34308	04521	70354	76945	87388
93	22878	76387	78369	66640	58239	30212	12720	09169	67017	60727
94	87500	80446	43063	65135	01941	90948	06700	24121	60740	80525
95	37339	54979	95043	12332	08323	02317	93679	19199	99301	97339
96	18954	28076	67347	88897	72946	70483	56727	84655	93493	98487
97	44158	30177	42091	17405	07523	08563	94731	47646	24983	90879
98	35915	53210	61857	76484	26845	62687	12560	66207	59259	84022
99	80455	39046	57587	98528	45554	63984	19246	61731	56068	26555
100	32653	75715	70737	62813	79557	30238	70151	43957	20127	79639

TABLE A2

Area under the normal distribution curve from z to infinity, calculated using IMSL subroutine MDNOR

z	0.00	0.01	0.02	0.03	0.04	0.05	0.06	0.07	0.08	0.09
0.0	0.5000	0.4960	0.4920	0.4880	0.4840	0.4801	0.4761	0.4721	0.4681	0.4641
0.1	0.4602	0.4562	0.4522	0.4483	0.4443	0.4404	0.4364	0.4325	0.4286	0.4247
0.2	0.4207	0.4168	0.4129	0.4090	0.4052	0.4013	0.3974	0.3936	0.3897	0.3859
0.3	0.3821	0.3783	0.3745	0.3707	0.3669	0.3632	0.3594	0.3557	0.3520	0.3483
0.4	0.3446	0.3409	0.3372	0.3336	0.3300	0.3264	0.3228	0.3192	0.3156	0.3121
0.5	0.3085	0.3050	0.3015	0.2981	0.2946	0.2912	0.2877	0.2843	0.2810	0.2776
0.6	0.2743	0.2709	0.2676	0.2643	0.2611	0.2578	0.2546	0.2514	0.2483	0.2451
0.7	0.2420	0.2389	0.2358	0.2327	0.2296	0.2266	0.2236	0.2206	0.2177	0.2148
0.8	0.2119	0.2090	0.2061	0.2033	0.2005	0.1977	0.1949	0.1922	0.1894	0.1867
0.9	0.1841	0.1814	0.1788	0.1762	0.1736	0.1711	0.1685	0.1660	0.1635	0.1611
1.0	0.1587	0.1562	0.1539	0.1515	0.1492	0.1469	0.1446	0.1423	0.1401	0.1379
1.1	0.1357	0.1335	0.1314	0.1292	0.1271	0.1251	0.1230	0.1210	0.1190	0.1170
1.2	0.1151	0.1131	0.1112	0.1093	0.1075	0.1056	0.1038	0.1020	0.1003	0.0985
1.3	0.0968	0.0951	0.0934	0.0918	0.0901	0.0885	0.0869	0.0853	0.0838	0.0823
1.4	0.0808	0.0793	0.0778	0.0764	0.0749	0.0735	0.0721	0.0708	0.0694	0.0681
1.5	0.0668	0.0655	0.0643	0.0630	0.0618	0.0606	0.0594	0.0582	0.0571	0.0559
1.6	0.0548	0.0537	0.0526	0.0516	0.0505	0.0495	0.0485	0.0475	0.0465	0.0455
1.7	0.0446	0.0436	0.0427	0.0418	0.0409	0.0401	0.0392	0.0384	0.0375	0.0367
1.8	0.0359	0.0351	0.0344	0.0336	0.0329	0.0322	0.0314	0.0307	0.0301	0.0294
1.9	0.0287	0.0281	0.0274	0.0268	0.0262	0.0256	0.0250	0.0244	0.0239	0.0233
2.0	0.0228	0.0222	0.0217	0.0212	0.0207	0.0202	0.0197	0.0192	0.0188	0.0183
2.1	0.0179	0.0174	0.0170	0.0166	0.0162	0.0158	0.0154	0.0150	0.0146	0.0143
2.2	0.0139	0.0136	0.0132	0.0129	0.0125	0.0122	0.0119	0.0116	0.0113	0.0110
2.3	0.0107	0.0104	0.0102	0.0099	0.0096	0.0094	0.0091	0.0089	0.0087	0.0084
2.4	0.0082	0.0080	0.0078	0.0075	0.0073	0.0071	0.0069	0.0068	0.0066	0.0064
2.5	.00621	.00604	.00587	.00570	.00554	.00539	.00523	.00508	.00494	.00480
2.6	.00466	.00453	.00440	.00427	.00415	.00402	.00391	.00379	.00368	.00357
2.7	.00347	.00336	.00326	.00317	.00307	.00298	.00289	.00280	.00272	.00264
2.8	.00256	.00248	.00240	.00233	.00226	.00219	.00212	.00205	.00199	.00193
2.9	.00187	.00181	.00175	.00169	.00164	.00159	.00154	.00149	.00144	.00139
3.0	.00135	.00131	.00126	.00122	.00118	.00114	.00111	.00107	.00104	.00100
3.1	.00097	.00094	.00090	.00087	.00084	.00082	.00079	.00076	.00074	.00071
3.2	.00069	.00066	.00064	.00062	.00060	.00058	.00056	.00054	.00052	.00050
3.3	.00048	.00047	.00045	.00043	.00042	.00040	.00039	.00038	.00036	.00035
3.4	.00034	.00032	.00031	.00030	.00029	.00028	.00027	.00026	.00025	.00024
3.5	.00023	.00022	.00022	.00021	.00020	.00019	.00019	.00018	.00017	.00017
3.6	.00016	.00015	.00015	.00014	.00014	.00013	.00013	.00012	.00012	.00011
3.7	.00011	.00010	.00010	.00010	.00009	.00009	.00008	.00008	.00008	.00008
3.8	.00007	.00007	.00007	.00006	.00006	.00006	.00006	.00005	.00005	.00005
3.9	.00005	.00005	.00004	.00004	.00004	.00004	.00004	.00004	.00003	.00003

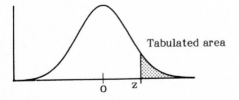

Tabulated area

TABLE A3

Critical values of the t distribution, calculated using IMSL subroutine MDSTI

D.f.	Level of significance, α							
	0.300	0.200	0.100	0.050	0.025	0.010	0.005	0.001
1	0.727	1.376	3.078	6.314	12.71	31.82	63.66	318.3
2	0.617	1.061	1.886	2.920	4.303	6.965	9.925	22.33
3	0.584	0.978	1.638	2.353	3.182	4.541	5.841	10.22
4	0.569	0.941	1.533	2.132	2.776	3.747	4.604	7.173
5	0.559	0.920	1.476	2.015	2.571	3.365	4.032	5.893
6	0.553	0.906	1.440	1.943	2.447	3.143	3.707	5.208
7	0.549	0.896	1.415	1.895	2.365	2.998	3.499	4.785
8	0.546	0.889	1.397	1.860	2.306	2.896	3.355	4.501
9	0.543	0.883	1.383	1.833	2.262	2.821	3.250	4.297
10	0.542	0.879	1.372	1.812	2.228	2.764	3.169	4.144
11	0.540	0.876	1.363	1.796	2.201	2.718	3.106	4.025
12	0.539	0.873	1.356	1.782	2.179	2.681	3.055	3.930
13	0.538	0.870	1.350	1.771	2.160	2.650	3.012	3.852
14	0.537	0.868	1.345	1.761	2.145	2.624	2.977	3.787
15	0.536	0.866	1.341	1.753	2.131	2.602	2.947	3.733
16	0.535	0.865	1.337	1.746	2.120	2.583	2.921	3.686
17	0.534	0.863	1.333	1.740	2.110	2.567	2.898	3.646
18	0.534	0.862	1.330	1.734	2.101	2.552	2.878	3.610
19	0.533	0.861	1.328	1.729	2.093	2.539	2.861	3.579
20	0.533	0.860	1.325	1.725	2.086	2.528	2.845	3.552
21	0.532	0.859	1.323	1.721	2.080	2.518	2.831	3.527
22	0.532	0.858	1.321	1.717	2.074	2.508	2.819	3.505
23	0.532	0.858	1.319	1.714	2.069	2.500	2.807	3.485
24	0.531	0.857	1.318	1.711	2.064	2.492	2.797	3.467
25	0.531	0.856	1.316	1.708	2.060	2.485	2.787	3.450
26	0.531	0.856	1.315	1.706	2.056	2.479	2.779	3.435
27	0.531	0.855	1.314	1.703	2.052	2.473	2.771	3.421
28	0.530	0.855	1.313	1.701	2.048	2.467	2.763	3.408
29	0.530	0.854	1.311	1.699	2.045	2.462	2.756	3.396
30	0.530	0.854	1.310	1.697	2.042	2.457	2.750	3.385
40	0.529	0.851	1.303	1.684	2.021	2.423	2.704	3.307
60	0.527	0.848	1.296	1.671	2.000	2.390	2.660	3.232
80	0.526	0.846	1.292	1.664	1.990	2.374	2.639	3.195
120	0.526	0.845	1.289	1.658	1.980	2.358	2.617	3.160
∞	0.524	0.842	1.282	1.645	1.960	2.326	2.576	3.090

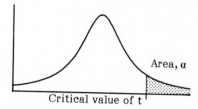

Area, α

Critical value of t

TABLE A4

Critical values for the F distribution for a level of significance of $\alpha = 0.05$, calculated using IMSL subroutine MDFI

		Degrees of freedom for numerator, n_1							
	1	2	3	4	5	6	7	8	9
1	161.	199.	216.	225.	230.	234.	237.	239.	241.
2	18.5	19.0	19.2	19.2	19.3	19.3	19.4	19.4	19.4
3	10.1	9.55	9.28	9.12	9.01	8.94	8.89	8.85	8.81
4	7.71	6.94	6.59	6.39	6.26	6.16	6.09	6.04	6.00
5	6.61	5.79	5.41	5.19	5.05	4.95	4.88	4.82	4.77
6	5.99	5.14	4.76	4.53	4.39	4.28	4.21	4.15	4.10
7	5.59	4.74	4.35	4.12	3.97	3.87	3.79	3.73	3.68
8	5.32	4.46	4.07	3.84	3.69	3.58	3.50	3.44	3.39
9	5.12	4.26	3.86	3.63	3.48	3.37	3.29	3.23	3.18
10	4.96	4.10	3.71	3.48	3.33	3.22	3.14	3.07	3.02
11	4.84	3.98	3.59	3.36	3.20	3.09	3.01	2.95	2.90
12	4.75	3.89	3.49	3.26	3.11	3.00	2.91	2.85	2.80
13	4.67	3.81	3.41	3.18	3.03	2.92	2.83	2.77	2.71
14	4.60	3.74	3.34	3.11	2.96	2.85	2.76	2.70	2.65
15	4.54	3.68	3.29	3.06	2.90	2.79	2.71	2.64	2.59
16	4.49	3.63	3.24	3.01	2.85	2.74	2.66	2.59	2.54
17	4.45	3.59	3.20	2.96	2.81	2.70	2.61	2.55	2.49
18	4.41	3.55	3.16	2.93	2.77	2.66	2.58	2.51	2.46
19	4.38	3.52	3.13	2.90	2.74	2.63	2.54	2.48	2.42
20	4.35	3.49	3.10	2.87	2.71	2.60	2.51	2.45	2.39
21	4.32	3.47	3.07	2.84	2.68	2.57	2.49	2.42	2.37
22	4.30	3.44	3.05	2.82	2.66	2.55	2.46	2.40	2.34
23	4.28	3.42	3.03	2.80	2.64	2.53	2.44	2.37	2.32
24	4.26	3.40	3.01	2.78	2.62	2.51	2.42	2.36	2.30
25	4.24	3.39	2.99	2.76	2.60	2.49	2.40	2.34	2.28
30	4.17	3.32	2.92	2.69	2.53	2.42	2.33	2.27	2.21
40	4.08	3.23	2.84	2.61	2.45	2.34	2.25	2.18	2.12
60	4.00	3.15	2.76	2.53	2.37	2.25	2.17	2.10	2.04
120	3.92	3.07	2.68	2.45	2.29	2.18	2.09	2.02	1.96
∞	3.84	3.00	2.60	2.37	2.21	2.10	2.01	1.94	1.88

Degrees of freedom for denominator, n_2

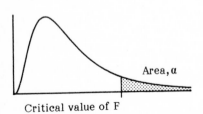

Area, α

Critical value of F

Degrees of freedom for numerator, n_1									
10	12	15	20	30	40	60	120	∞	
242.	244.	246.	248.	250.	251.	252.	253.	254.	1
19.4	19.4	19.4	19.4	19.5	19.5	19.5	19.5	19.5	2
8.79	8.74	8.70	8.66	8.62	8.59	8.57	8.55	8.53	3
5.96	5.91	5.86	5.80	5.75	5.72	5.69	5.66	5.63	4
4.74	4.68	4.62	4.56	4.50	4.46	4.43	4.40	4.37	5
4.06	4.00	3.94	3.87	3.81	3.77	3.74	3.70	3.67	6
3.64	3.57	3.51	3.44	3.38	3.34	3.30	3.27	3.23	7
3.35	3.28	3.22	3.15	3.08	3.04	3.01	2.97	2.93	8
3.14	3.07	3.01	2.94	2.86	2.83	2.79	2.75	2.71	9
2.98	2.91	2.85	2.77	2.70	2.66	2.62	2.58	2.54	10
2.85	2.79	2.72	2.65	2.57	2.53	2.49	2.45	2.40	11
2.75	2.69	2.62	2.54	2.47	2.43	2.38	2.34	2.30	12
2.67	2.60	2.53	2.46	2.38	2.34	2.30	2.25	2.21	13
2.60	2.53	2.46	2.39	2.31	2.27	2.22	2.18	2.13	14
2.54	2.48	2.40	2.33	2.25	2.20	2.16	2.11	2.07	15
2.49	2.42	2.35	2.28	2.19	2.15	2.11	2.06	2.01	16
2.45	2.38	2.31	2.23	2.15	2.10	2.06	2.01	1.96	17
2.41	2.34	2.27	2.19	2.11	2.06	2.02	1.97	1.92	18
2.38	2.31	2.23	2.16	2.07	2.03	1.98	1.93	1.88	19
2.35	2.28	2.20	2.12	2.04	1.99	1.95	1.90	1.84	20
2.32	2.25	2.18	2.10	2.01	1.96	1.92	1.87	1.81	21
2.30	2.23	2.15	2.07	1.98	1.94	1.89	1.84	1.78	22
2.27	2.20	2.13	2.05	1.91	1.86	1.81	1.76		23
2.25	2.18	2.11	2.03	1.94	1.89	1.84	1.79	1.73	24
2.24	2.16	2.09	2.01	1.92	1.87	1.82	1.77	1.71	25
2.16	2.09	2.01	1.93	1.84	1.79	1.74	1.68	1.62	30
2.08	2.00	1.92	1.84	1.74	1.69	1.64	1.58	1.51	40
1.99	1.92	1.84	1.75	1.65	1.59	1.53	1.47	1.39	60
1.91	1.83	1.75	1.66	1.55	1.50	1.43	1.35	1.25	120
1.83	1.75	1.67	1.57	1.46	1.39	1.32	1.22	1.00	∞

Degrees of freedom for denominator, n_2

TABLE A4 (Continued)

Critical values for the F distribution for a level of significance of $\alpha = 0.01$, calculated using IMSL subroutine MDFI

		Degrees of freedom for numerator, n_1							
	1	2	3	4	5	6	7	8	9
1	4052.	4999.	5403.	5625.	5764.	5859.	5928.	5981.	6022.
2	98.5	99.0	99.2	99.2	99.3	99.3	99.4	99.4	99.4
3	34.1	30.8	29.5	28.7	28.2	27.9	27.7	27.5	27.3
4	21.2	18.0	16.7	16.0	15.5	15.2	15.0	14.8	14.7
5	16.3	13.3	12.1	11.4	11.0	10.7	10.5	10.3	10.2
6	13.7	10.9	9.78	9.15	8.75	8.47	8.26	8.10	7.98
7	12.2	9.55	8.45	7.85	7.46	7.19	6.99	6.84	6.72
8	11.3	8.65	7.59	7.01	6.63	6.37	6.18	6.03	5.91
9	10.6	8.02	6.99	6.42	6.06	5.80	5.61	5.47	5.35
10	10.0	7.56	6.55	5.99	5.64	5.39	5.20	5.06	4.94
11	9.65	7.21	6.22	5.67	5.32	5.07	4.89	4.74	4.63
12	9.33	6.93	5.95	5.41	5.06	4.82	4.64	4.50	4.39
13	9.07	6.70	5.74	5.21	4.86	4.62	4.44	4.30	4.19
14	8.86	6.51	5.56	5.04	4.69	4.46	4.28	4.14	4.03
15	8.68	6.36	5.42	4.89	4.56	4.32	4.14	4.00	3.89
16	8.53	6.23	5.29	4.77	4.44	4.20	4.03	3.89	3.78
17	8.40	6.11	5.18	4.67	4.34	4.10	3.93	3.79	3.68
18	8.29	6.01	5.09	4.58	4.25	4.01	3.84	3.71	3.60
19	8.18	5.93	5.01	4.50	4.17	3.94	3.77	3.63	3.52
20	8.10	5.85	4.94	4.43	4.10	3.87	3.70	3.56	3.46
21	8.02	5.78	4.87	4.37	4.04	3.81	3.64	3.51	3.40
22	7.95	5.72	4.82	4.31	3.99	3.76	3.59	3.45	3.35
23	7.88	5.66	4.76	4.26	3.94	3.71	3.54	3.41	3.30
24	7.82	5.61	4.72	4.22	3.90	3.67	3.50	3.36	3.26
25	7.77	5.57	4.68	4.18	3.85	3.63	3.46	3.32	3.22
30	7.56	5.39	4.51	4.02	3.70	3.47	3.30	3.17	3.07
40	7.31	5.18	4.31	3.83	3.51	3.29	3.12	2.99	2.89
60	7.08	4.98	4.13	3.65	3.34	3.12	2.95	2.82	2.72
120	6.85	4.79	3.95	3.48	3.17	2.96	2.79	2.66	2.56
∞	6.63	4.61	3.78	3.32	3.02	2.80	2.64	2.51	2.41

Degrees of freedom for denominator, n_2

Degrees of freedom for numerator, n_1									
10	12	15	20	30	40	60	120	∞	
6056.	6106.	6157.	6209.	6261.	6287.	6313.	6339.	6366.	1
99.4	99.4	99.4	99.4	99.5	99.5	99.5	99.5	99.5	2
27.2	27.1	26.9	26.7	26.5	26.4	26.3	26.2	26.1	3
14.5	14.4	14.2	14.0	13.8	13.7	13.7	13.6	13.5	4
10.1	9.89	9.72	9.55	9.38	9.29	9.20	9.11	9.02	5
7.87	7.72	7.56	7.40	7.23	7.14	7.06	6.97	6.88	6
6.62	6.47	6.31	6.16	5.99	5.91	5.82	5.74	5.65	7
5.81	5.67	5.52	5.36	5.20	5.12	5.03	4.95	4.86	8
5.26	5.11	4.96	4.81	4.65	4.57	4.48	4.40	4.31	9
4.85	4.71	4.56	4.41	4.25	4.17	4.08	4.00	3.91	10
4.54	4.40	4.25	4.10	3.94	3.86	3.78	3.69	3.60	11
4.30	4.16	4.01	3.86	3.70	3.62	3.54	3.45	3.36	12
4.10	3.96	3.82	3.66	3.51	3.43	3.34	3.25	3.17	13
3.94	3.80	3.66	3.51	3.35	3.27	3.18	3.09	3.00	14
3.80	3.67	3.52	3.37	3.21	3.13	3.05	2.96	2.87	15
3.69	3.55	3.41	3.26	3.10	3.02	2.93	2.84	2.75	16
3.59	3.46	3.31	3.16	3.00	2.92	2.83	2.75	2.65	17
3.51	3.37	3.23	3.08	2.92	2.84	2.75	2.66	2.57	18
3.43	3.30	3.15	3.00	2.84	2.76	2.67	2.58	2.49	19
3.37	3.23	3.09	2.94	2.78	2.69	2.61	2.52	2.42	20
3.31	3.17	3.03	2.88	2.72	2.64	2.55	2.46	2.36	21
3.26	3.12	2.98	2.83	2.67	2.58	2.50	2.40	2.31	22
3.21	3.07	2.93	2.78	2.62	2.54	2.45	2.35	2.26	23
3.17	3.03	2.89	2.74	2.58	2.49	2.40	2.31	2.21	24
3.13	2.99	2.85	2.70	2.54	2.45	2.36	2.27	2.17	25
2.98	2.84	2.70	2.55	2.39	2.30	2.21	2.11	2.01	30
2.80	2.66	2.52	2.37	2.20	2.11	2.02	1.92	1.80	40
2.63	2.50	2.35	2.20	2.03	1.94	1.84	1.73	1.60	60
2.47	2.34	2.19	2.03	1.86	1.76	1.66	1.53	1.38	120
2.32	2.18	2.04	1.88	1.70	1.59	1.47	1.32	1.00	∞

Degrees of freedom for denominator, n_2

TABLE A5

Critical values of the Chi-square distribution, calculated using IMSL subroutine MDCHI

	Level of significance, α									
D.f.	0.90	0.85	0.80	0.75	0.70	0.65	0.60	0.55	0.50	0.45
1	0.0158	0.0358	0.0642	0.102	0.148	0.206	0.275	0.357	0.455	0.571
2	0.211	0.325	0.446	0.575	0.713	0.861	1.02	1.20	1.39	1.60
3	0.584	0.798	1.01	1.21	1.42	1.64	1.87	2.11	2.37	2.64
4	1.06	1.37	1.65	1.92	2.19	2.47	2.75	3.05	3.36	3.69
5	1.61	1.99	2.34	2.67	3.00	3.33	3.66	4.00	4.35	4.73
6	2.20	2.66	3.07	3.45	3.83	4.20	4.57	4.95	5.35	5.77
7	2.83	3.36	3.82	4.25	4.67	5.08	5.49	5.91	6.35	6.80
8	3.49	4.08	4.59	5.07	5.53	5.98	6.42	6.88	7.34	7.83
9	4.17	4.82	5.38	5.90	6.39	6.88	7.36	7.84	8.34	8.86
10	4.86	5.57	6.18	6.74	7.27	7.78	8.29	8.81	9.34	9.89
11	5.58	6.34	6.99	7.58	8.15	8.70	9.24	9.78	10.3	10.9
12	6.30	7.11	7.81	8.44	9.03	9.61	10.2	10.8	11.3	11.9
13	7.04	7.90	8.63	9.30	9.93	10.5	11.1	11.7	12.3	13.0
14	7.79	8.69	9.47	10.2	10.8	11.5	12.1	12.7	13.3	14.0
15	8.55	9.50	10.3	11.0	11.7	12.4	13.0	13.7	14.3	15.0
16	9.31	10.3	11.2	11.9	12.6	13.3	14.0	14.7	15.3	16.0
17	10.1	11.1	12.0	12.8	13.5	14.2	14.9	15.6	16.3	17.1
18	10.9	11.9	12.9	13.7	14.4	15.2	15.9	16.6	17.3	18.1
19	11.7	12.8	13.7	14.6	15.4	16.1	16.9	17.6	18.3	19.1
20	12.4	13.6	14.6	15.5	16.3	17.0	17.8	18.6	19.3	20.1
21	13.2	14.4	15.4	16.3	17.2	18.0	18.8	19.5	20.3	21.1
22	14.0	15.3	16.3	17.2	18.1	18.9	19.7	20.5	21.3	22.2
23	14.8	16.1	17.2	18.1	19.0	19.9	20.7	21.5	22.3	23.2
24	15.7	17.0	18.1	19.0	19.9	20.8	21.7	22.5	23.3	24.2
25	16.5	17.8	18.9	19.9	20.9	21.8	22.6	23.5	24.3	25.2
26	17.3	18.7	19.8	20.8	21.8	22.7	23.6	24.5	25.3	26.2
27	18.1	19.5	20.7	21.7	22.7	23.6	24.5	25.4	26.3	27.3
28	18.9	20.4	21.6	22.7	23.6	24.6	25.5	26.4	27.3	28.3
29	19.8	21.2	22.5	23.6	24.6	25.5	26.5	27.4	28.3	29.3
30	20.6	22.1	23.4	24.5	25.5	26.5	27.4	28.4	29.3	30.3
40	29.1	30.9	32.4	33.7	34.9	36.0	37.1	38.2	39.3	40.5
60	46.5	48.8	50.6	52.3	53.8	55.2	56.6	58.0	59.3	60.7
80	64.3	67.0	69.2	71.1	72.9	74.6	76.2	77.8	79.3	80.9
100	82.4	85.4	88.0	90.1	92.1	94.0	95.8	97.6	99.3	101.
120	101.	104.	107.	109.	111.	113.	115.	117.	119.	121.

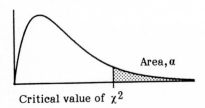

Area, α

Critical value of χ^2

0.40	0.35	0.30	0.25	0.20	0.10	Level of significance, α 0.05	0.025	0.01	0.005	D.f.
0.708	0.874	1.07	1.32	1.64	2.71	3.84	5.02	6.64	7.90	1
1.83	2.10	2.41	2.77	3.22	4.60	5.99	7.38	9.22	10.6	2
2.95	3.28	3.67	4.11	4.64	6.25	7.82	9.36	11.3	12.8	3
4.04	4.44	4.88	5.39	5.99	7.78	9.49	11.2	13.3	14.8	4
5.13	5.57	6.06	6.63	7.29	9.24	11.1	12.8	15.1	16.8	5
6.21	6.70	7.23	7.84	8.56	10.6	12.6	14.5	16.8	18.5	6
7.28	7.81	8.38	9.04	9.80	12.0	14.1	16.0	18.5	20.3	7
8.35	8.91	9.52	10.2	11.0	13.4	15.5	17.5	20.1	21.9	8
9.41	10.0	10.7	11.4	12.2	14.7	16.9	19.0	21.7	23.6	9
10.5	11.1	11.8	12.5	13.4	16.0	18.3	20.5	23.2	25.2	10
11.5	12.2	12.9	13.7	14.6	17.3	19.7	21.9	24.8	26.7	11
12.6	13.3	14.0	14.8	15.8	18.6	21.0	23.3	26.2	28.2	12
13.6	14.3	15.1	16.0	17.0	19.8	22.4	24.7	27.7	29.9	13
14.7	15.4	16.2	17.1	18.2	21.1	23.7	26.1	29.2	31.4	14
15.7	16.5	17.3	18.2	19.3	22.3	25.0	27.5	30.6	32.9	15
16.8	17.6	18.4	19.4	20.5	23.5	26.3	28.9	32.0	34.3	16
17.8	18.6	19.5	20.5	21.6	24.8	27.6	30.2	33.4	35.8	17
18.9	19.7	20.6	21.6	22.8	26.0	28.9	31.5	34.8	37.2	18
19.9	20.8	21.7	22.7	23.9	27.2	30.1	32.9	36.2	38.6	19
21.0	21.8	22.8	23.8	25.0	28.4	31.4	34.2	37.6	40.0	20
22.0	22.9	23.9	24.9	26.2	29.6	32.7	35.5	39.0	41.4	21
23.0	23.9	24.9	26.0	27.3	30.8	33.9	36.8	40.3	42.8	22
24.1	25.0	26.0	27.1	28.4	32.0	35.2	38.1	41.7	44.2	23
25.1	26.1	27.1	28.2	29.6	33.2	36.4	39.4	43.0	45.6	24
26.1	27.1	28.2	29.3	30.7	34.4	37.7	40.7	44.3	47.0	25
27.2	28.2	29.2	30.4	31.8	35.6	38.9	41.9	45.7	48.3	26
28.2	29.2	30.3	31.5	32.9	36.7	40.1	43.2	47.0	49.7	27
29.2	30.3	31.4	32.6	34.0	37.9	41.3	44.5	48.3	51.0	28
30.3	31.3	32.5	33.7	35.1	39.1	42.6	45.7	49.6	52.4	29
31.3	32.4	33.5	34.8	36.3	40.3	43.8	47.0	50.9	53.7	30
41.6	42.8	44.2	45.6	47.3	51.8	55.8	59.3	63.7	66.8	40
62.1	63.6	65.2	67.0	69.0	74.4	79.1	83.3	88.4	92.0	60
82.6	84.3	86.1	88.1	90.4	96.6	102.	107.	112.	116.	80
103.	105.	107.	109.	112.	118.	124.	130.	136.	140.	100
123.	125.	128.	130.	133.	140.	147.	152.	159.	164.	120

TABLE A6

Critical values of the correlation coefficient, r

D.f.	Level of significance, α					
	0.100	0.050	0.025	0.010	0.005	0.001
1	0.951	0.988	0.997	1.000	1.000	1.000
2	0.800	0.900	0.950	0.980	0.990	0.999
3	0.687	0.805	0.878	0.934	0.959	0.991
4	0.608	0.729	0.811	0.882	0.917	0.974
5	0.551	0.669	0.754	0.833	0.875	0.951
6	0.507	0.621	0.707	0.789	0.834	0.925
7	0.472	0.582	0.666	0.750	0.798	0.898
8	0.443	0.549	0.632	0.715	0.765	0.872
9	0.419	0.521	0.602	0.685	0.735	0.847
10	0.398	0.497	0.576	0.658	0.708	0.823
11	0.380	0.476	0.553	0.634	0.684	0.801
12	0.365	0.458	0.532	0.612	0.661	0.780
13	0.351	0.441	0.514	0.592	0.641	0.760
14	0.338	0.426	0.497	0.574	0.623	0.742
15	0.327	0.412	0.482	0.558	0.606	0.725
16	0.317	0.400	0.468	0.543	0.590	0.708
17	0.308	0.389	0.456	0.529	0.575	0.693
18	0.299	0.378	0.444	0.516	0.561	0.679
19	0.291	0.369	0.433	0.503	0.549	0.665
20	0.284	0.360	0.423	0.492	0.537	0.652
21	0.277	0.352	0.413	0.482	0.526	0.640
22	0.271	0.344	0.404	0.472	0.515	0.629
23	0.265	0.337	0.396	0.462	0.505	0.618
24	0.260	0.330	0.388	0.453	0.496	0.607
25	0.255	0.323	0.381	0.445	0.487	0.597
26	0.250	0.317	0.374	0.437	0.479	0.588
27	0.245	0.311	0.367	0.430	0.471	0.579
28	0.241	0.306	0.361	0.423	0.463	0.570
29	0.237	0.301	0.355	0.416	0.456	0.562
30	0.233	0.296	0.349	0.409	0.449	0.554
40	0.202	0.257	0.304	0.358	0.393	0.490
60	0.165	0.211	0.250	0.295	0.325	0.408
80	0.143	0.183	0.217	0.257	0.283	0.357
100	0.128	0.164	0.195	0.230	0.254	0.321
200	0.091	0.116	0.138	0.164	0.181	0.230
400	0.064	0.082	0.098	0.116	0.128	0.164
600	0.052	0.067	0.080	0.095	0.105	0.134
800	0.045	0.058	0.069	0.082	0.091	0.116
1000	0.041	0.052	0.062	0.073	0.081	0.104

TABLE A7

Conversion of correlation coefficient r to z using equation 4.9

r	0.00	0.01	0.02	0.03	0.04	0.05	0.06	0.07	0.08	0.09
0.0	0.0000	0.0100	0.0200	0.0300	0.0400	0.0500	0.0601	0.0701	0.0802	0.0902
0.1	0.1003	0.1104	0.1206	0.1307	0.1409	0.1511	0.1614	0.1717	0.1820	0.1923
0.2	0.2027	0.2132	0.2237	0.2342	0.2448	0.2554	0.2661	0.2769	0.2877	0.2986
0.3	0.3095	0.3205	0.3316	0.3428	0.3541	0.3654	0.3769	0.3884	0.4001	0.4118
0.4	0.4236	0.4356	0.4477	0.4599	0.4722	0.4847	0.4973	0.5101	0.5230	0.5361
0.5	0.5493	0.5627	0.5763	0.5901	0.6042	0.6184	0.6328	0.6475	0.6625	0.6777
0.6	0.6931	0.7089	0.7250	0.7414	0.7582	0.7753	0.7928	0.8107	0.8291	0.8480
0.7	0.8673	0.8872	0.9076	0.9287	0.9505	0.9730	0.9962	1.0203	1.0454	1.0714
0.8	1.0986	1.1270	1.1568	1.1881	1.2212	1.2562	1.2933	1.3331	1.3758	1.4219
0.9	1.4722	1.4775	1.4828	1.4882	1.4937	1.4992	1.5047	1.5103	1.5160	1.5217

r	0.000	0.001	0.002	0.003	0.004	0.005	0.006	0.007	0.008	0.009
0.91	1.5275	1.5334	1.5393	1.5453	1.5513	1.5574	1.5636	1.5698	1.5762	1.5826
0.92	1.5890	1.5956	1.6022	1.6089	1.6157	1.6226	1.6296	1.6366	1.6438	1.6510
0.93	1.6584	1.6658	1.6734	1.6811	1.6888	1.6967	1.7047	1.7129	1.7211	1.7295
0.94	1.7380	1.7467	1.7555	1.7645	1.7736	1.7828	1.7923	1.8019	1.8117	1.8216
0.95	1.8318	1.8421	1.8527	1.8635	1.8745	1.8857	1.8972	1.9090	1.9210	1.9333
0.96	1.9459	1.9588	1.9721	1.9857	1.9996	2.0139	2.0287	2.0439	2.0595	2.0756
0.97	2.0923	2.1095	2.1273	2.1457	2.1649	2.1847	2.2054	2.2269	2.2494	2.2729
0.98	2.2976	2.3235	2.3507	2.3796	2.4101	2.4427	2.4774	2.5147	2.5550	2.5987
0.99	2.6467	2.6996	2.7587	2.8257	2.9031	2.9945	3.1063	3.2504	3.4534	3.8002

TABLE A8

Atomic weights based on $C^{12} = 12$, taken from Weast (1974, p.B1)

Ag	107.868	Eu	151.96	Mo	95.94	Se	78.96
Al	26.98154	F	18.9984	N	14.0067	Si	28.086
Ar	39.948	Fe	55.847	Na	22.98977	Sm	150.4
As	74.9216	Ga	69.72	Nb	92.9064	Sn	118.69
Au	196.9665	Gd	157.25	Nd	144.24	Sr	87.62
B	10.81	Ge	72.59	Ne	20.179	Ta	180.9479
Ba	137.34	H	1.0079	Ni	58.71	Tb	158.9254
Be	9.01218	He	4.0026	O	15.9994	Te	127.6
Bi	208.9808	Hf	178.49	Os	190.2	Th	232.0381
Br	79.904	Hg	200.59	P	30.97376	Ti	47.9
C	12.011	Ho	164.9303	Pb	207.2	Tl	204.37
Ca	40.08	I	126.9045	Pd	106.4	Tm	168.9342
Cd	112.4	In	114.82	Pr	140.9077	U	238.029
Ce	140.12	Ir	192.22	Pt	195.09	V	50.9414
Cl	35.453	K	39.098	Rb	85.4678	W	183.85
Co	58.9332	Kr	83.8	Re	186.2	Xe	131.3
Cr	51.996	La	138.9055	Rh	102.9055	Y	88.9059
Cs	132.9054	Li	6.941	Ru	101.07	Yb	173.04
Cu	63.546	Lu	174.97	S	32.06	Zn	65.38
Dy	162.5	Mg	24.305	Sb	121.75	Zr	91.22
Er	167.26	Mn	54.938	Sc	44.9559		

TABLE A9

Molecular weights of some oxides and factors to convert them to elements by weight, based on the atomic weights of Table A8.

Oxide	Mol.wt.	Factor	Oxide	Mol.wt.	Factor
Al_2O_3	101.96128	0.529251	MnO_2	86.9368	0.631930
B_2O_3	69.6182	0.310551	Mn_3O_4	228.8116	0.720304
BaO	153.3394	0.895660	Na_2O	61.97894	0.741857
BeO	25.01158	0.360320	Nb_2O_5	265.8098	0.699044
CO_2	44.0098	0.272916	Nd_2O_3	336.4782	0.857351
CaO	56.0794	0.714701	NiO	74.7094	0.785845
CeO_2	172.1188	0.814089	P_2O_5	141.94452	0.436421
Ce_2O_3	328.2382	0.853770	PbO	223.1994	0.928318
CoO	74.9326	0.786483	Pr_2O_3	329.8136	0.854469
Cr_2O_3	151.9902	0.684202	Rb_2O	186.9350	0.914412
Cs_2O	281.8102	0.943226	SO_3	80.0582	0.400459
CuO	79.5454	0.798865	Sb_2O_3	291.4982	0.835340
Dy_2O_3	372.9982	0.871318	Sc_2O_3	137.9100	0.651960
Er_2O_3	382.5182	0.874520	SiO_2	60.0848	0.467439
EuO	167.9594	0.904742	Sm_2O_3	348.7982	0.862390
Eu_2O_3	351.9182	0.863610	SnO_2	150.6888	0.787650
FeO	71.8464	0.777311	SrO	103.6194	0.845595
Fe_2O_3	159.6922	0.699433	Ta_2O_5	441.8928	0.818967
Ga_2O_3	187.4382	0.743925	Tb_2O_3	365.8490	0.868803
Gd_2O_3	362.4982	0.867591	ThO_2	264.0369	0.878809
GeO_2	104.5888	0.694051	TiO_2	79.8988	0.599508
H_2O	18.0152	0.111894	Ti_2O_3	143.7982	0.666211
HfO_2	210.4888	0.847979	Tm_2O_3	385.8666	0.875609
Ho_2O_3	377.8588	0.872973	UO_2	270.0278	0.881498
K_2O	94.1954	0.830147	U_3O_8	842.0822	0.848002
La_2O_3	325.8092	0.852680	V_2O_5	181.8798	0.560166
Li_2O	29.8814	0.464570	Y_2O_3	225.8100	0.787440
Lu_2O_3	397.9382	0.879383	Yb_2O_3	394.0782	0.878201
MgO	40.3044	0.603036	ZnO	81.3794	0.803397
MnO	70.9374	0.774457	ZrO_2	123.2188	0.740309

TABLE A10

Molecular weights of some sulphides and factors to convert them to elements by weight, based on the atomic weights of Table A8.

Sulphide	Mol.wt.	Factor	Sulphide	Mol.wt.	Factor
Ag_2S	247.796	0.870619	NiS	90.77	0.646800
As_2S_3	246.0232	0.609061	OsS_3	254.32	0.747877
Bi_2S_3	514.1416	0.812931	PbS	239.26	0.866004
CdS	144.46	0.778070	PdS	138.46	0.768453
CuS	95.606	0.664665	PtS	227.15	0.858860
Cu_2S	159.152	0.798557	PtS_2	259.21	0.752633
FeS	87.907	0.635296	RhS_2	167.0255	0.616107
FeS_2	119.967	0.465520	RuS_2	165.19	0.611841
HgS	232.65	0.862196	Sb_2S_3	339.68	0.716851
IrS_2	256.34	0.749863	SnS_2	182.81	0.649253
Ir_2S_3	480.62	0.799883	Tl_2S_3	504.92	0.809514
MnS	86.998	0.631486	ZnS	97.44	0.670977
MoS_2	160.06	0.599400			

TABLE A11

Conversion factors to change weights of some oxides and sulphides from one redox state to another, based on the atomic weights of Table A8.

Multiply	by	to obtain
CeO_2	0.953522	Ce_2O_3
EuO	1.047629	Eu_2O_3
FeO	1.111344	Fe_2O_3
FeO	1.074230	Fe_3O_4
Fe_2O_3	0.966604	Fe_3O_4
FeS	1.364704	FeS_2
IrS_2	0.937466	Ir_2S_3
MnO	1.225543	MnO_2
MnO	1.075181	Mn_3O_4
MnO_2	0.877310	Mn_3O_4
PtS	1.141140	PtS_2
TiO_2	0.899877	Ti_2O_3
UO_2	1.039501	U_3O_8

TABLE A12

Alphabetical list of the ideal mineral molecules whose compositions are given in Table A13. The formulae are based mainly on those given by Deer et al. (1966) and the molecular weights were calculated using the atomic weights of Table A8 and subroutines from the CLAIR data system library. The minerals are numbered according to the order in which they occur in Table A13.

		Mol. wt.	Formula
61.	Acmite	231.00517	$NaFeSi_2O_6$
91.	Aenigmatite	861.61854	$Na_2Fe_5TiSi_6O_{20}$
42.	Åkermanite	272.6328	$Ca_2MgSi_2O_7$
106.	Albite	262.22451	$NaAlSi_3O_8$
16.	Almandine	497.75488	$Fe_3Al_2Si_3O_{12}$
102.	Amesite	557.37056	$Mg_4Al_2(Al_2Si_2O_{10})(OH)_8$
120.	Analcite	220.15491	$NaAlSi_2O_6.H_2O$
19.	Andradite	508.1848	$Ca_3Fe_2Si_3O_{12}$
145.	Anhydrite	136.1376	$CaSO_4$
107.	Anorthite	278.21028	$CaAl_2Si_2O_8$
90.	Arfvedsonite	930.02825	$Na_3Fe_4AlSi_8O_{22}(OH)_2$
54.	Axinite(Ca)	554.35538	$Ca_3Al_2BSi_4O_{15}.OH$
55.	Axinite(Fe)	601.65638	$Fe_3Al_2BSi_4O_{15}.OH$
142.	Baryte	233.3976	$BaSO_4$
45.	Beryl	537.50482	$Be_3Al_2Si_6O_{18}$
95.	Biotite	511.88714	$KFe_3AlSi_3O_{10}(OH)_2$
138.	Brucite	58.3196	$Mg(OH)_2$
146.	Calcite	100.0892	$CaCO_3$
117.	Cancrinite	958.31820	$6NaAlSiO_4.Na_2CO_3$
143.	Celestine	183.6776	$SrSO_4$
108.	Celsian	375.47028	$BaAl_2Si_2O_8$
103.	Chamosite	683.53856	$Fe_4Al_2(Al_2Si_2O_{10})(OH)_8$
155.	Chlorapatite	520.76708	$Ca_5(PO_4)_3Cl$
30.	Chloritoid	503.81536	$Fe_2Al_4Si_2O_{10}(OH)_4$
7.	Chondrodite	339.7068	$Mg(OH)_2.2Mg_2SiO_4$
8.	Chondrodite(F)	343.6890	$Mg(F)_2.2Mg_2SiO_4$
137.	Chromite	223.8366	$FeCr_2O_4$
99.	Clinochlore	555.79848	$Mg_5Al(AlSi_3O_{10})(OH)_8$
11.	Clinohumite	621.0940	$Mg(OH)_2.4Mg_2SiO_4$
12.	Clinohumite(F)	625.0762	$Mg(F)_2.4Mg_2SiO_4$
47.	Cordierite(Fe)	648.03936	$Fe_2Al_4Si_5O_{18}$
46.	Cordierite(Mg)	584.95536	$Mg_2Al_4Si_5O_{18}$
122.	Corundum	101.96128	Al_2O_3
67.	Cummingtonite	780.8244	$Mg_7Si_8O_{22}(OH)_2$
140.	Diaspore	59.98824	$AlO(OH)$
58.	Diopside	216.5534	$CaMgSi_2O_6$

		Mol. wt.	Formula
150.	Dolomite	184.4034	$CaMg(CO_3)_2$
48.	Dravite	958.75301	$NaMg_3Al_6B_3Si_6O_{27}(OH)_4$
49.	Dravite(F)	966.71741	$NaMg_3Al_6B_3Si_6O_{27}(F)_4$
89.	Eckermanite	803.86025	$Na_3Mg_4AlSi_8O_{22}(OH)_2$
73.	Edenite	834.25971	$NaCa_2Mg_5AlSi_7O_{22}(OH)_2$
52.	Elbaite	1873.44364	$Na_2Li_3Al_3Al_{12}B_6Si_{12}O_{54}(OH)_8$
53.	Elbaite(F)	1889.37244	$Na_2Li_3Al_3Al_{12}B_6Si_{12}O_{54}(F)_8$
56.	Enstatite	100.3892	$MgSiO_3$
36.	Epidote	483.22818	$Ca_2FeAl_2Si_3O_{12}(OH)$
2.	Fayalite	203.7776	Fe_2SiO_4
72.	Ferroactinolite	970.0844	$Ca_2Fe_5Si_8O_{22}(OH)_2$
151.	Ferrodolomite	215.9454	$CaFe(CO_3)_2$
74.	Ferroedenite	991.96971	$NaCa_2Fe_5AlSi_7O_{22}(OH)_2$
70.	Ferrogedrite	941.67856	$Fe_5Al_2Al_2Si_6O_{22}(OH)_2$
82.	Ferroglaucophane	878.17302	$Na_2Fe_3Al_2Si_8O_{22}(OH)_2$
78.	Ferropargasite	961.99979	$NaCa_2Fe_4AlAl_2Si_6O_{22}(OH)_2$
86.	Ferrorichterite	975.98394	$Na_2CaFe_5Si_8O_{22}(OH)_2$
57.	Ferrosilite	131.9312	$FeSiO_3$
76.	Ferrotschermakite	910.14456	$Ca_2Fe_3Al_2Al_2Si_6O_{22}(OH)_2$
158.	Fluorite	78.0768	CaF_2
154.	Fluor-apatite	504.31248	$Ca_5(PO_4)_3F$
1.	Forsterite	140.6936	Mg_2SiO_4
134.	Franklinite	241.0716	$ZnFe_2O_4$
129.	Gahnite	183.34068	$ZnAl_2O_4$
130.	Galaxite	172.89868	$MnAl_2O_4$
69.	Gedrite	783.96856	$Mg_5Al_2Al_2Si_6O_{22}(OH)_2$
41.	Gehlenite	274.20488	$Ca_2Al_2SiO_7$
139.	Gibbsite	78.00344	$Al(OH)_3$
81.	Glaucophane	783.54702	$Na_2Mg_3Al_2Si_8O_{22}(OH)_2$
141.	Goethite	88.8537	$FeO(OH)$
101.	Greenalite	743.4784	$Fe_6Si_4O_{10}(OH)_8$
18.	Grossular	450.45388	$Ca_3Al_2Si_3O_{12}$
68.	Grunerite	1001.6184	$Fe_7Si_8O_{22}(OH)_2$
144.	Gypsum	172.1680	$CaSO_4.2H_2O$
159.	Halite	58.44277	$NaCl$
80.	Hastingsite	990.86525	$NaCa_2Fe_4FeAl_2Si_6O_{22}(OH)_2$
116.	Haüyne	1941.53688	$6CaAl_2Si_2O_8.2CaSO_4$
59.	Hedenbergite	248.0954	$CaFeSi_2O_6$
123.	Hematite	159.6922	Fe_2O_3
128.	Hercynite	173.80768	$FeAl_2O_4$
9.	Humite	480.4004	$Mg(OH)_2.3Mg_2SiO_4$
10.	Humite(F)	484.3826	$Mg(F)_2.3Mg_2SiO_4$

TABLE A12 (continued)

		Mol. wt.	Formula
156.	Hydroxyapatite	502.32138	$Ca_5(PO_4)_3OH$
124.	Ilmenite	151.7452	$FeTiO_3$
135.	Jacobsite	230.6296	$MnFe_2O_4$
62.	Jadeite	202.13971	$NaAlSi_2O_6$
60.	Johannsenite	247.1864	$CaMnSi_2O_6$
112.	Kalsilite	158.16314	$KAlSiO_4$
104.	Kaolin	258.16128	$Al_2Si_2O_5(OH)_4$
88.	Katophorite	974.87948	$Na_2CaFe_4FeAlSi_7O_{22}(OH)_2$
32.	Larnite	172.2436	Ca_2SiO_4
39.	Lawsonite	314.24068	$CaAl_2Si_2O_7(OH)_2.H_2O$
113.	Leucite	218.24794	$KAlSi_2O_6$
136.	Magnesiochromite	192.2946	$MgCr_2O_4$
131.	Magnesioferrite	199.9966	$MgFe_2O_4$
79.	Magnesiohastingsite	864.69725	$NaCa_2Mg_4FeAl_2Si_6O_{22}(OH)_2$
87.	Magnesiokatophorite	848.71148	$Na_2CaMg_4FeAlSi_7O_{22}(OH)_2$
83.	Magnesioriebeckite	841.27794	$Na_2Mg_3Fe_2Si_8O_{22}(OH)_2$
147.	Magnesite	84.3142	$MgCO_3$
132.	Magnetite	231.5386	Fe_3O_4
96.	Margarite	398.18676	$CaAl_2Al_2Si_2O_{10}(OH)_2$
118.	Marialite	320.66728	$NaAlSi_3O_8.NaCl$
119.	Meionite	378.29948	$CaAl_2Si_2O_8.CaCO_3$
33.	Merwinite	328.7122	$Ca_3MgSi_2O_8$
157.	Monazite	235.09136	$CePO_4$
4.	Monticellite	156.4686	$CaMgSiO_4$
26.	Mullite	426.05344	$Al_6Si_2O_{13}$
92.	Muscovite	398.30922	$KAl_3Si_3O_{10}(OH)_2$
110.	Nepheline	142.05491	$NaAlSiO_4$
111.	Nepheline(Na,K)	584.32787	$Na_3KAl_4Si_4O_{16}$
5.	Norbergite	199.0132	$Mg(OH)_2.Mg_2SiO_4$
6.	Norbergite(F)	202.9954	$Mg(F)_2.Mg_2SiO_4$
115.	Nosean	994.36660	$6NaAlSiO_4.Na_2SO_4$
105.	Orthoclase	278.33274	$KAlSi_3O_8$
93.	Paragonite	382.20099	$NaAl_3Si_3O_{10}(OH)_2$
77.	Pargasite	835.83179	$NaCa_2Mg_4AlAl_2Si_6O_{22}(OH)_2$
66.	Pectolite	332.41027	$Ca_2NaSi_3O_8(OH)$
121.	Periclase	40.3044	MgO
125.	Perovskite	135.9782	$CaTiO_3$
37.	Piemontite($MnAl_2$)	482.31918	$Ca_2MnAl_2Si_3O_{12}(OH)$
38.	Piemontite(Mn_2Al)	510.27564	$Ca_2Mn_2AlSi_3O_{12}(OH)$

	Mol. wt.	Formula
94. Phlogopite	417.26114	$KMg_3AlSi_3O_{10}(OH)_2$
40. Pumpellyite	943.08720	$Ca_4MgAl_5Si_6O_{23}(OH)_3.2H_2O$
15. Pyrope	403.12888	$Mg_3Al_2Si_3O_{12}$
97. Pyrophyllite	180.15784	$AlSi_2O_5(OH)$
109. Quartz	60.0848	SiO_2
43. Rankinite	288.4078	$Ca_3Si_2O_7$
85. Richterite	818.27394	$Na_2CaMg_5Si_8O_{22}(OH)_2$
84. Riebeckite	935.90394	$Na_2Fe_3Fe_2Si_8O_{22}(OH)_2$
149. Rhodochrosite	114.9472	$MnCO_3$
65. Rhodonite	131.0222	$MnSiO_3$
126. Rutile	79.8988	TiO_2
31. Sapphirine	344.61616	$Mg_2Al_4SiO_{10}$
50. Schorl	1053.37901	$NaFe_3Al_6B_3Si_6O_{27}(OH)_4$
51. Schorl(F)	1061.34341	$NaFe_3Al_6B_3Si_6O_{27}(F)_4$
100. Serpentine	554.2264	$Mg_6Si_4O_{10}(OH)_8$
148. Siderite	115.8562	$FeCO_3$
25. Sillimanite	162.04608	Al_2SiO_5
114. Sodalite	969.21500	$6NaAlSiO_4.2NaCl$
17. Spessartine	495.02788	$Mn_3Al_2Si_3O_{12}$
14. Sphene	196.0630	$CaTiSiO_5$
127. Spinel	142.26568	$MgAl_2O_4$
63. Spodumene	186.09094	$LiAlSi_2O_6$
34. Spurrite	444.5764	$2Ca_2SiO_4.CaCO_3$
29. Staurolite	851.86536	$Fe_2Al_9Si_4O_{23}(OH)$
152. Strontianite	147.6292	$SrCO_3$
98. Talc	379.2676	$Mg_3Si_4O_{10}(OH)_2$
3. Tephroite	201.9596	Mn_2SiO_4
44. Tilleyite	488.5862	$Ca_3Si_2O_7.2CaCO_3$
27. Topaz	180.06128	$Al_2SiO_4(OH)_2$
28. Topaz(F)	184.04348	$Al_2SiO_4(F)_2$
71. Tremolite	812.3744	$Ca_2Mg_5Si_8O_{22}(OH)_2$
75. Tschermakite	815.51856	$Ca_2Mg_3Al_2Al_2Si_6O_{22}(OH)_2$
133. Ulvöspinel	223.5916	Fe_2TiO_4
20. Uvarovite	500.4828	$Ca_3Cr_2Si_3O_{12}$
22. Vesuvianite(Fe)	1485.20296	$Ca_{10}Fe_2Al_4Si_9O_{34}(OH)_4$
24. Vesuvianite(Fe,F)	1493.16736	$Ca_{10}Fe_2Al_4Si_9O_{34}(F)_4$
21. Vesuvianite(Mg)	1422.11896	$Ca_{10}Mg_2Al_4Si_9O_{34}(OH)_4$
23. Vesuvianite(Mg,F)	1430.08336	$Ca_{10}Mg_2Al_4Si_9O_{34}(F)_4$
153. Witherite	197.3492	$BaCO_3$
64. Wollastonite	116.1642	$CaSiO_3$
13. Zircon	183.3036	$ZrSiO_4$
35. Zoisite	454.36272	$Ca_2Al_3Si_3O_{12}(OH)$

TABLE A13

The theoretical compositions of some ideal mineral molecules, calculated using the atomic weights of Table A8 and subroutines from the CLAIR data system library. The order in which the minerals are presented is basically the same as that used in Deer et al. (1966). Table A12 may be used as an alphabetical index to the mineral names in this Table. Ox.eq is the oxygen equivalent.

		SiO_2	FeO	MgO	MnO	CaO
1.	Forsterite	42.71	–	57.29	–	–
2.	Fayalite	29.49	70.51	–	–	–
3.	Tephroite	29.75	–	–	70.25	–
4.	Monticellite	38.40	–	25.76	–	35.84

		SiO_2	MgO	H_2O	F	Ox.eq
5.	Norbergite	30.19	60.76	9.05	–	–
6.	Norbergite(F)	29.60	59.56	–	18.72	–7.88
7.	Chondrodite	35.37	59.32	5.30	–	–
8.	Chondrodite(F)	34.96	58.63	–	11.06	–4.66
9.	Humite	37.52	58.73	3.75	–	–
10.	Humite(F)	37.21	58.25	–	7.84	–3.30
11.	Clinohumite	38.70	58.40	2.90	–	–
12.	Clinohumite(F)	38.45	58.03	–	6.08	–2.56

		SiO_2	TiO_2	CaO	ZrO_2
13.	Zircon	32.78	–	–	67.22
14.	Sphene	30.65	40.75	28.60	–

		SiO_2	Al_2O_3	Fe_2O_3	FeO	MnO	MgO	CaO	Cr_2O_3
15.	Pyrope	44.71	25.29	–	–	–	29.99	–	–
16.	Almandine	36.21	20.48	–	43.30	–	–	–	–
17.	Spessartine	36.41	20.60	–	–	42.99	–	–	–
18.	Grossular	40.02	22.64	–	–	–	–	37.35	–
19.	Andradite	35.47	–	31.42	–	–	–	33.11	–
20.	Uvarovite	36.02	–	–	–	–	–	33.62	30.37

	SiO$_2$	Al$_2$O$_3$	FeO	MgO	CaO	H$_2$O	F	Ox.eq
21. Vesuvianite(Mg)	38.03	14.34	–	5.67	39.43	2.53	–	–
22. Vesuvianite(Fe)	36.41	13.73	9.67	–	37.76	2.43	–	–
23. Vesuvianite(Mg,F)	37.81	14.26	–	5.64	39.21	–	5.31	-2.24
24. Vesuvianite(Fe,F)	36.22	13.66	9.62	–	37.56	–	5.09	-2.14
25. Sillimanite	37.08	62.92	–	–	–	–	–	–
26. Mullite	28.21	71.79	–	–	–	–	–	–
27. Topaz	33.37	56.63	–	–	–	10.01	–	–
28. Topaz(F)	32.65	55.40	–	–	–	–	20.65	-8.69
29. Staurolite	28.21	53.86	16.87	–	–	1.06	–	–
30. Chloritoid	23.85	40.48	28.52	–	–	7.15	–	–
31. Sapphirine	17.44	59.17	–	23.39	–	–	–	–

	SiO$_2$	MgO	CaO	CO$_2$
32. Larnite	34.88	–	65.12	–
33. Merwinite	36.56	12.26	51.18	–
34. Spurrite	27.03	–	63.07	9.90

	SiO$_2$	Al$_2$O$_3$	Fe$_2$O$_3$	Mn$_2$O$_3$	MgO	CaO	H$_2$O
35. Zoisite	39.67	33.66	–	–	–	24.68	1.98
36. Epidote	37.30	21.10	16.52	–	–	23.21	1.86
37. Piemontite(MnAl$_2$)	37.37	21.14	–	16.37	–	23.25	1.87
38. Piemontite(Mn$_2$Al)	35.32	9.99	–	30.94	–	21.98	1.77
39. Lawsonite	38.24	32.45	–	–	–	17.85	11.47
40. Pumpellyite	38.23	27.03	–	–	4.27	23.79	6.69

	SiO$_2$	Al$_2$O$_3$	MgO	CaO	CO$_2$
41. Gehlenite	21.91	37.18	–	40.90	–
42. Åkermanite	44.08	–	14.78	41.14	–
43. Rankinite	41.67	–	–	58.33	–
44. Tilleyite	24.60	–	–	57.39	18.02

TABLE A13 (continued)

	SiO_2	Al_2O_3	FeO	MgO	BeO
45. Beryl	67.07	18.97	–	–	13.96
46. Cordierite(Mg)	51.36	34.86	–	13.78	–
47. Cordierite(Fe)	46.36	31.47	22.17	–	–

	SiO_2	Al_2O_3	FeO	MgO	Na_2O	Li_2O	B_2O_3	H_2O	F	Ox.eq
48. Dravite	37.60	31.90	–	12.61	3.23	–	10.89	3.76	–	–
49. Dravite(F)	37.29	31.64	–	12.51	3.21	–	10.80	–	7.86	–3.31
50. Schorl	34.22	29.04	20.46	–	2.94	–	9.91	3.42	–	–
51. Schorl(F)	33.97	28.82	20.31	–	2.92	–	9.84	–	7.16	–3.01
52. Elbaite	38.49	40.82	–	–	3.31	2.39	11.15	3.85	–	–
53. Elbaite(F)	38.16	40.47	–	–	3.28	2.37	11.05	–	8.04	–3.39

	SiO_2	Al_2O_3	FeO	CaO	B_2O_3	H_2O
54. Axinite(Ca)	43.35	18.39	–	30.35	6.28	1.62
55. Axinite(Fe)	39.95	16.95	35.82	–	5.79	1.50

	SiO_2	Al_2O_3	Fe_2O_3	FeO	MnO	MgO	CaO	Na_2O	Li_2O
56. Enstatite	59.85	–	–	–	–	40.15	–	–	–
57. Ferrosilite	45.54	–	–	54.46	–	–	–	–	–
58. Diopside	55.49	–	–	–	–	18.61	25.90	–	–
59. Hedenbergite	48.44	–	–	28.96	–	–	22.60	–	–
60. Johannsenite	48.61	–	–	–	28.70	–	22.69	–	–
61. Acmite	52.02	–	34.56	–	–	–	–	13.42	–
62. Jadeite	59.45	25.22	–	–	–	–	–	15.33	–
63. Spodumene	64.58	27.40	–	–	–	–	–	–	8.03
64. Wollastonite	51.72	–	–	–	–	–	48.28	–	–
65. Rhodonite	45.86	–	–	–	54.14	–	–	–	–

	SiO_2	CaO	Na_2O	H_2O
66. Pectolite	54.23	33.74	9.32	2.71

		SiO_2	Al_2O_3	Fe_2O_3	FeO	MgO	CaO	Na_2O	H_2O
67.	Cummingtonite	61.56	–	–	–	36.13	–	–	2.31
68.	Grunerite	47.99	–	–	50.21	–	–	–	1.80
69.	Gedrite	45.99	26.01	–	–	25.71	–	–	2.30
70.	Ferrogedrite	38.28	21.66	–	38.15	–	–	–	1.91
71.	Tremolite	59.17	–	–	–	24.81	13.81	–	2.22
72.	Ferroactinolite	49.55	–	–	37.03	–	11.56	–	1.86
73.	Edenite	50.42	6.11	–	–	24.16	13.44	3.71	2.16
74.	Ferroedenite	42.40	5.14	–	36.21	–	11.31	3.12	1.82
75.	Tschermakite	44.21	25.01	–	–	14.83	13.75	–	2.21
76.	Ferrotschermakite	39.61	22.41	–	23.68	–	12.32	–	1.98
77.	Pargasite	43.13	18.30	–	–	19.29	13.42	3.71	2.16
78.	Ferropargasite	37.47	15.90	–	29.87	–	11.66	3.22	1.87
79.	Magnesiohastingsite	41.69	11.79	9.23	–	18.64	12.97	3.58	2.08
80.	Hastingsite	36.38	10.29	8.06	29.00	–	11.32	3.13	1.82
81.	Glaucophane	61.35	13.01	–	–	15.43	–	7.91	2.30
82.	Ferroglaucophane	54.74	11.61	–	24.54	–	–	7.06	2.05
83.	Magnesioriebeckite	57.14	–	18.98	–	14.37	–	7.37	2.14
84.	Riebeckite	51.36	–	17.06	23.03	–	–	6.62	1.92
85.	Richterite	58.74	–	–	–	24.63	6.85	7.57	2.20
86.	Ferrorichterite	49.25	–	–	36.81	–	5.75	6.35	1.85
87.	Magnesiokatophorite	49.56	6.01	9.41	–	19.00	6.61	7.30	2.12
88.	Katophorite	43.14	5.23	8.19	29.48	–	5.75	6.36	1.85
89.	Eckermanite	59.80	6.34	–	–	20.06	–	11.57	2.24
90.	Arfvedsonite	51.68	5.48	–	30.90	–	–	10.00	1.94

	SiO_2	TiO_2	FeO	Na_2O
91. Aenigmatite	41.84	9.27	41.69	7.19

		SiO_2	Al_2O_3	FeO	MgO	CaO	Na_2O	K_2O	H_2O
92.	Muscovite	45.25	38.40	–	–	–	–	11.82	4.52
93.	Paragonite	47.16	40.02	–	–	–	8.11	–	4.71
94.	Phlogopite	43.20	12.22	–	28.98	–	–	11.29	4.32
95.	Biotite	35.21	9.96	42.11	–	–	–	9.20	3.52
96.	Margarite	30.18	51.21	–	–	14.08	–	–	4.52
97.	Pyrophyllite	66.70	28.30	–	–	–	–	–	5.00
98.	Talc	63.37	–	–	31.88	–	–	–	4.75
99.	Clinochlore	32.43	18.35	–	36.26	–	–	–	12.97
100.	Serpentine	43.36	–	–	43.63	–	–	–	13.00
101.	Greenalite	32.33	–	57.98	–	–	–	–	9.69
102.	Amesite	21.56	36.59	–	28.92	–	–	–	12.93
103.	Chamosite	17.58	29.83	42.04	–	–	–	–	10.54
104.	Kaolin	46.55	39.50	–	–	–	–	–	13.96

TABLE A13 (continued)

	SiO$_2$	Al$_2$O$_3$	CaO	Na$_2$O	K$_2$O	BaO
105. Orthoclase	64.76	18.32	–	–	16.92	–
106. Albite	68.74	19.44	–	11.82	–	–
107. Anorthite	43.19	36.65	20.16	–	–	–
108. Celsian	32.01	27.16	–	–	–	40.84
109. Quartz	100.00	–	–	–	–	–
110. Nepheline	42.30	35.89	–	21.82	–	–
111. Nepheline(Na,K)	41.13	34.90	–	15.91	8.06	–
112. Kalsilite	37.99	32.23	–	–	29.78	–
113. Leucite	55.06	23.36	–	–	21.58	–

	SiO$_2$	Al$_2$O$_3$	CaO	Na$_2$O	Cl	SO$_3$	CO$_2$	H$_2$O	Ox.eq
114. Sodalite	37.20	31.56	–	25.58	7.32	–	–	–	-1.65
115. Nosean	36.26	30.76	–	24.93	–	8.05	–	–	–
116. Haüyne	37.14	31.51	23.11	–	–	8.25	–	–	–
117. Cancrinite	37.62	31.92	–	25.87	–	–	4.59	–	–
118. Marialite	56.21	15.90	–	19.33	11.06	–	–	–	-2.49
119. Meionite	31.77	26.95	29.65	–	–	–	11.63	–	–
120. Analcite	54.58	23.16	–	14.08	–	–	–	8.18	–

	TiO$_2$	Al$_2$O$_3$	Fe$_2$O$_3$	FeO	MgO	CaO
121. Periclase	–	–	–	–	100.00	–
122. Corundum	–	100.00	–	–	–	–
123. Hematite	–	–	100.00	–	–	–
124. Ilmenite	52.65	–	–	47.35	–	–
125. Perovskite	58.76	–	–	–	–	41.24
126. Rutile	100.00	–	–	–	–	–

	TiO$_2$	Al$_2$O$_3$	Fe$_2$O$_3$	FeO	MnO	MgO	ZnO	Cr$_2$O$_3$
127. Spinel	–	71.67	–	–	–	28.33	–	–
128. Hercynite	–	58.66	–	41.34	–	–	–	–
129. Gahnite	–	55.61	–	–	–	–	44.39	–
130. Galaxite	–	58.97	–	–	41.03	–	–	–
131. Magnesioferrite	–	–	79.85	–	–	20.15	–	–
132. Magnetite	–	–	68.97	31.03	–	–	–	–
133. Ulvöspinel	35.73	–	–	64.27	–	–	–	–
134. Franklinite	–	–	66.24	–	–	–	33.76	–
135. Jacobsite	–	–	69.24	–	30.76	–	–	–
136. Magnesiochromite	–	–	–	–	–	20.96	–	79.04
137. Chromite	–	–	–	32.10	–	–	–	67.90

	Al$_2$O$_3$	Fe$_2$O$_3$	MgO	H$_2$O
138. Brucite	–	–	69.11	30.89
139. Gibbsite	65.36	–	–	34.64
140. Diaspore	84.98	–	–	15.02
141. Goethite	–	89.86	–	10.14

	CaO	BaO	SrO	SO$_3$	H$_2$O
142. Baryte	–	65.70	–	34.30	–
143. Celestine	–	–	56.41	43.59	–
144. Gypsum	32.57	–	–	46.50	20.93
145. Anhydrite	41.19	–	–	58.81	–

	FeO	MnO	MgO	CaO	SrO	BaO	CO$_2$
146. Calcite	–	–	–	56.03	–	–	43.97
147. Magnesite	–	–	47.80	–	–	–	52.20
148. Siderite	62.01	–	–	–	–	–	37.99
149. Rhodochrosite	–	61.71	–	–	–	–	38.29
150. Dolomite	–	–	21.86	30.41	–	–	47.73
151. Ferrodolomite	33.27	–	–	25.97	–	–	40.76
152. Strontianite	–	–	–	–	70.19	–	29.81
153. Witherite	–	–	–	–	–	77.70	22.30

	CaO	P$_2$O$_5$	Ce$_2$O$_3$	F	Cl	H$_2$O	Ox.eq
154. Fluor-apatite	55.60	42.22	–	3.77	–	–	-1.59
155. Chlorapatite	53.84	40.89	–	–	6.81	–	-1.54
156. Hydroxyapatite	55.82	42.39	–	–	–	1.79	–
157. Monazite	–	30.19	69.81	–	–	–	–

	CaO	Na$_2$O	F	Cl	Ox.eq
158. Fluorite	71.83	–	48.67	–	-20.49
159. Halite	–	53.03	–	60.66	-13.69

TABLE A14

Partial analyses of the silicates of Table A13 arranged in order of increasing weight percent SiO_2.

		SiO_2	Al_2O_3	FeO	MgO	CaO	Na_2O	K_2O
31.	Sapphirine	17.44	59.17	–	23.39	–	–	–
103.	Chamosite	17.58	29.83	42.04	–	–	–	–
102.	Amesite	21.56	36.59	–	28.92	–	–	–
41.	Gehlenite	21.91	37.18	–	–	40.90	–	–
30.	Chloritoid	23.85	40.48	28.52	–	–	–	–
44.	Tilleyite	24.60	–	–	–	57.39	–	–
34.	Spurrite	27.03	–	–	–	63.07	–	–
26.	Mullite	28.21	71.79	–	–	–	–	–
29.	Staurolite	28.21	53.86	16.87	–	–	–	–
2.	Fayalite	29.49	–	70.51	–	–	–	–
6.	Norbergite(F)	29.60	–	–	59.56	–	–	–
3.	Tephroite	29.75	–	–	–	–	–	–
96.	Margarite	30.18	51.21	–	–	14.08	–	–
5.	Norbergite	30.19	–	–	60.76	–	–	–
14.	Sphene	30.65	–	–	–	28.60	–	–
119.	Meionite	31.77	26.95	–	–	29.65	–	–
108.	Celsian	32.01	27.16	–	–	–	–	–
101.	Greenalite	32.33	–	57.98	–	–	–	–
99.	Clinochlore	32.43	18.35	–	36.26	–	–	–
28.	Topaz(F)	32.65	55.40	–	–	–	–	–
13.	Zircon	32.78	–	–	–	–	–	–
27.	Topaz	33.37	56.63	–	–	–	–	–
51.	Schorl(F)	33.97	28.82	20.31	–	–	2.92	–
50.	Schorl	34.22	29.04	20.46	–	–	2.94	–
32.	Larnite	34.88	–	–	–	65.12	–	–
8.	Chondrodite(F)	34.96	–	–	58.63	–	–	–
95.	Biotite	35.21	9.96	42.11	–	–	–	9.20
38.	Piemontite(Mn_2Al)	35.32	9.99	–	–	21.98	–	–
7.	Chondrodite	35.37	–	–	59.32	–	–	–
19.	Andradite	35.47	–	–	–	33.11	–	–
20.	Uvarovite	36.02	–	–	–	33.62	–	–
16.	Almandine	36.21	20.48	43.30	–	–	–	–
24.	Vesuvianite(Fe,F)	36.22	13.66	9.62	–	37.56	–	–
115.	Nosean	36.26	30.76	–	–	–	24.93	–
80.	Hastingsite	36.38	10.29	29.00	–	11.32	3.13	–
22.	Vesuvianite(Fe)	36.41	13.73	9.67	–	37.76	–	–
17.	Spessartine	36.41	20.60	–	–	–	–	–
33.	Merwinite	36.56	–	–	12.26	51.18	–	–

		SiO$_2$	Al$_2$O$_3$	FeO	MgO	CaO	Na$_2$O	K$_2$O
25.	Sillimanite	37.08	62.92	–	–	–	–	–
116.	Haüyne	37.14	31.51	–	–	23.11	–	–
114.	Sodalite	37.20	31.56	–	–	–	25.58	–
10.	Humite(F)	37.21	–	–	58.25	–	–	–
49.	Dravite(F)	37.29	31.64	–	12.51	–	3.21	–
36.	Epidote	37.30	21.10	–	–	23.21	–	–
37.	Piemontite(MnAl$_2$)	37.37	21.14	–	–	23.25	–	–
78.	Ferropargasite	37.47	15.90	29.87	–	11.66	3.22	–
9.	Humite	37.52	–	–	58.73	–	–	–
48.	Dravite	37.60	31.90	–	12.61	–	3.23	–
117.	Cancrinite	37.62	31.92	–	–	–	25.87	–
23.	Vesuvianite(Mg,F)	37.81	14.26	–	5.64	39.21	–	–
112.	Kalsilite	37.99	32.23	–	–	–	–	29.78
21.	Vesuvianite(Mg)	38.03	14.34	–	5.67	39.43	–	–
53.	Elbaite(F)	38.16	40.47	–	–	–	3.28	–
40.	Pumpellyite	38.23	27.03	–	4.27	23.79	–	–
39.	Lawsonite	38.24	32.45	–	–	17.85	–	–
70.	Ferrogedrite	38.28	21.66	38.15	–	–	–	–
4.	Monticellite	38.40	–	–	25.76	35.84	–	–
12.	Clinohumite(F)	38.45	–	–	58.03	–	–	–
52.	Elbaite	38.49	40.82	–	–	–	3.31	–
11.	Clinohumite	38.70	–	–	58.40	–	–	–
76.	Ferrotschermakite	39.61	22.41	23.68	–	12.32	–	–
35.	Zoisite	39.67	33.66	–	–	24.68	–	–
55.	Axinite(Fe)	39.95	16.95	35.82	–	–	–	–
18.	Grossular	40.02	22.64	–	–	37.35	–	–
111.	Nepheline(Na,K)	41.13	34.90	–	–	–	15.91	8.06
43.	Rankinite	41.67	–	–	–	58.33	–	–
79.	Magnesiohastingsite	41.69	11.79	–	18.64	12.97	3.58	–
91.	Aenigmatite	41.84	–	41.69	–	–	7.19	–
110.	Nepheline	42.30	35.89	–	–	–	21.82	–
74.	Ferroedenite	42.40	5.14	36.21	–	11.31	3.12	–
1.	Forsterite	42.71	–	–	57.29	–	–	–
77.	Pargasite	43.13	18.30	–	19.29	13.42	3.71	–
88.	Katophorite	43.14	5.23	29.48	–	5.75	6.36	–
107.	Anorthite	43.19	36.65	–	–	20.16	–	–
94.	Phlogopite	43.20	12.22	–	28.98	–	–	11.29
54.	Axinite(Ca)	43.35	18.39	–	–	30.35	–	–
100.	Serpentine	43.36	–	–	43.63	–	–	–
42.	Åkermanite	44.08	–	–	14.78	41.14	–	–
75.	Tschermakite	44.21	25.01	–	14.83	13.75	–	–
15.	Pyrope	44.71	25.29	–	29.99	–	–	–

TABLE A14 (continued)

		SiO$_2$	Al$_2$O$_3$	FeO	MgO	CaO	Na$_2$O	K$_2$O
92.	Muscovite	45.25	38.40	-	-	-	-	11.82
57.	Ferrosilite	45.54	-	54.46	-	-	-	-
65.	Rhodonite	45.86	-	-	-	-	-	-
69.	Gedrite	45.99	26.01	-	25.71	-	-	-
47.	Cordierite(Fe)	46.36	31.47	22.17	-	-	-	-
104.	Kaolin	46.55	39.50	-	-	-	-	-
93.	Paragonite	47.16	40.02	-	-	-	8.11	-
68.	Grunerite	47.99	-	50.21	-	-	-	-
59.	Hedenbergite	48.44	-	28.96	-	22.60	-	-
60.	Johannsenite	48.61	-	-	-	22.69	-	-
86.	Ferrorichterite	49.25	-	36.81	-	5.75	6.35	-
72.	Ferroactinolite	49.55	-	37.03	-	11.56	-	-
87.	Magnesiokatophorite	49.56	6.01	-	19.00	6.61	7.30	-
73.	Edenite	50.42	6.11	-	24.16	13.44	3.71	-
46.	Cordierite(Mg)	51.36	34.86	-	13.78	-	-	-
84.	Riebeckite	51.36	-	23.03	-	-	6.62	-
90.	Arfvedsonite	51.68	5.48	30.90	-	-	10.00	-
64.	Wollastonite	51.72	-	-	-	48.28	-	-
61.	Acmite	52.02	-	-	-	-	13.42	-
66.	Pectolite	54.23	-	-	-	33.74	9.32	-
120.	Analcite	54.58	23.16	-	-	-	14.08	-
82.	Ferroglaucophane	54.74	11.61	24.54	-	-	7.06	-
113.	Leucite	55.06	23.36	-	-	-	-	21.58
58.	Diopside	55.49	-	-	18.61	25.90	-	-
118.	Marialite	56.21	15.90	-	-	-	19.33	-
83.	Magnesioriebeckite	57.14	-	-	14.37	-	7.37	-
85.	Richterite	58.74	-	-	24.63	6.85	7.57	-
71.	Tremolite	59.17	-	-	24.81	13.81	-	-
62.	Jadeite	59.45	25.22	-	-	-	15.33	-
89.	Eckermanite	59.80	6.34	-	20.06	-	11.57	-
56.	Enstatite	59.85	-	-	40.15	-	-	-
81.	Glaucophane	61.35	13.01	-	15.43	-	7.91	-
67.	Cummingtonite	61.56	-	-	36.13	-	-	-
98.	Talc	63.37	-	-	31.88	-	-	-
63.	Spodumene	64.58	27.40	-	-	-	-	-
105.	Orthoclase	64.76	18.32	-	-	-	-	16.92
97.	Pyrophyllite	66.70	28.30	-	-	-	-	-
45.	Beryl	67.07	18.97	-	-	-	-	-
106.	Albite	68.74	19.44	-	-	-	11.82	-
109.	Quartz	100.00	-	-	-	-	-	-

AUTHOR INDEX

SUBJECT INDEX